# Herpetology Book

OVER 40 YEARS OF
CHRONOLOGICAL EVENTS OF

## Charley

THE BOX TURTLE AND HIS
REHABILITATION AND
VETERINARY CARE

by

**Jeanne Hyland**

ISBN 979-8-9907147-2-4

Editing by Cortni Merritt at SRD Editing Services

Interior layout and cover art by Heidi Sutherlin at My Creative Pursuits

# Acknowledgments

I want to thank Fred and Vonda Stanley who made an extra copy of the only copy of this manuscript and for being so caring.

Thank you, Fred and Terri Baumbach, for all your help. Terri typed up my handwritten copy of this book.

Thank you to Patricia Humpreys for researching and her dedication to making this possible.

Thanks to Yohan Cielto for transferring my photos into the book.

Thank you, Janine and Marie Martin for organizing, editing, and finishing my book.

I would also like to thank the following persons for their invaluable contributions to this publication: John L. Behler, F. Wayne King, Tess Cook, J. G. Walls, Lenny Flank Jr., G. R. Zug, L. J. Vitt, J. P. Caldwell, Douglas R. Madar, Mike Pingleton, P. D. Vosjoli, R. J. Klingenberg, Amanda Ebenhack, Liz Palika; C. K. Dodd Jr., J. Patterson, R. D. Bartlett, Patricia Bartlett, and B. C. Case.

HYLAND

# Dedication

I first want to dedicate my book to the Lord who is the Holy One of Israel.

This book is dedicated in memory of my loving parents, Edward H. Hyland and Dorothy C. Hyland, who were with me when I purchased Charley in a pet store. My mom gave me the idea to write this book about Charley and always encouraged me not to give up on it. When Charley was sick, she reminded me, "Don't forget the book!"

This book is dedicated to Dr. John Charos, DVM, COO, for his patience and dedication to the preservation of Charley and others like him. His expertise has been invaluable, and his passion and love for the animals have shown he is a phenomenal veterinarian. I couldn't have thanked him enough for being such an amazing person in taking good care of Charley.

I would also like to recognize Dr. Robert A. Monaco, DVM, DABVP, who is a marvelous vet. I appreciated his suggestion in helping me with some of the ideas for this book. Dr. Monaco is outstanding in his knowledge and unselfish in his dedication to his patients and clients. He has also shown that he would risk his own safety for his patients.

Look for Charley throughout this book!

HYLAND

# Table of Contents

# A Turtle Is Not a Toy

Before purchasing or adopting any turtle or tortoise, you must know they are long-lived pets. They require daily maintenance, including soaking, nutrition, and checking on lighting and heat temperature. You must first find an exotic vet before purchasing any animal. The initial exam will be for parasites and the animal's overall health before taking it home. If one is unwilling to read through care books and learn the correct husbandry, save the turtle's life by leaving it in the store and saving your money. A turtle is not a toy for a child but a living creature. A turtle deserves respect like any other pet.

I am glad you are interested in Herpetology. May our grand Creator bless you and your family.

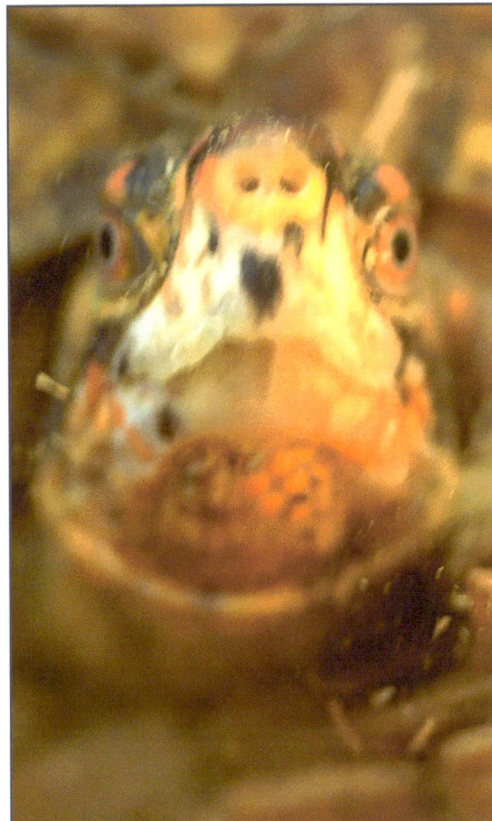

# Introduction

This book is a unique combination of the chronological events of several chelonians. However, the purpose of the book is primarily about the rehabilitation of one particular box turtle, Charley, who had four life-threatening illnesses in one year. The story details the process of correcting his husbandry, bringing the reader through the journey from sickness to wellness and correcting the conditions that caused it all in the first place. Veterinarians and clients can both take this journey to wellness with Charley, while experiencing his charming, calm character. It is written for anyone who has found a friend in a herp or a cherished chelonian like Charley and who wants to continue to keep a healthy, happy herp.

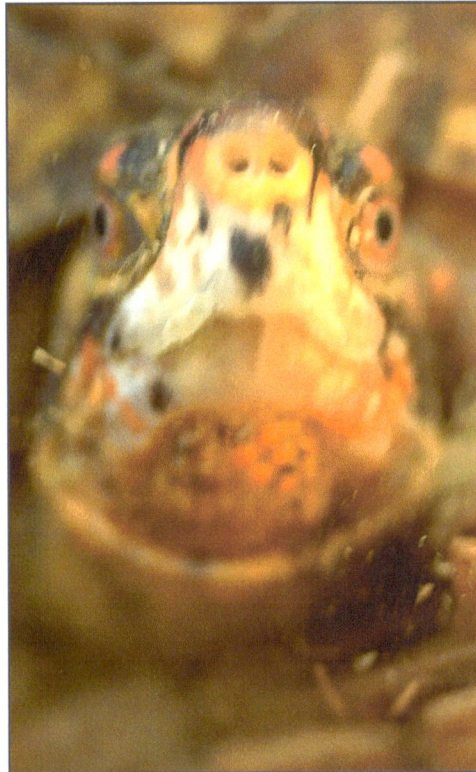

# Chapter 1

# Herpetology and My Herps

HYLAND

Herpetology is the scientific study of reptiles and amphibians. *Herp* is a shortened word for a reptile. A herp is a reptile, while a *herper* is a reptile owner. *Chelonians* are the scientific name for the turtle and tortoise species. Their taxonomy classification—common and scientific—follows. Below is Charley: Common box turtle or Eastern box turtle, subspecies: three-toed ***Terrapene carolina triunguis***.

Celie (below) is my first female three-toed box turtle, ***Terrapene carolina triunguis***. On September 9, 2015, a vet technician from the Animal Clinic left a message saying that he had found a girl for Charley. A little while later, in the office lobby, I met her and was very grateful. I took her home and named her Celie.

Charleen (below) is my second female three-toed box turtle, *Terrapene carolina triunguis*. On May 19, 2016, upon getting permission from the curator at Chelonian Research Institute to adopt the only three-toed box turtle they had, I said, "Thank you very much. This box turtle is just in time for my birthday."

Charlotte (below) is a subspecies of *Terrapene carolina bauri*, commonly known as the Florida box turtle. On July 8, 2013, my friend Heidi and I went to our local Pet Bazaar and met an employee, who gave me three-year-old Charlotte. He bred these turtles, and Charlotte was hatched in his house.

On April 27, 2003, I welcomed Boxy the Chinese box turtle for five years (*Cuora flavomarginata*). He was a huge herp but a gentle giant. Unfortunately, Boxy died while hibernating after I gave him to a vet technician in 2008.

Boxy (below):

Eddy (below): red-eared slider (**Trachemys scripta elegans**). I had Eddy for three years and donated him to Turtle Home in the Florida Keys. I found out he had shell rot. As an aquatic turtle, Eddy needed a lot of water to change. It was like taking care of a big fish and too much work for me.

Jeanne Hyland

I purchased two red-foot tortoises in 2008 and 2010, at Reptile World in Centereach, New York, a Long Island pet store. Edwina (on the left) and Rosie (on the right below) are *Chelonoidis carbonaria*. However, Rosie was a dwarf type known as a "cherry head." She was larger than most males or about the same size as a large male. Rosie weighed 12 pounds.

On January 2, 2017, I got Henry from the Chelonian Research Institute at five years old. He is my red-foot tortoise, *Chelonoidis carbonaria*.

I also got Terry from the Chelonian Research Institute on January 26, 2017, when she was four. She and Henry had been together for a few years, so I adopted them to remain together. I took care of Henry and Terry as a volunteer at the Institute. They had more turtles than they could care for at that time.

# Classification Chart

Biologists divide plants and animals into major groupings called phyla. The phylum Chordata includes all fishes, amphibians, reptiles, birds, and mammals. Phyla are divided into classes. Amphibians belong to the class Amphibia, reptiles to Reptilia. Classes are subdivided into orders, orders into families, families into genera (singular: genus), and genera into species (singular: species).

The species is the basic unit of our classification system and is generally what people have in mind when they talk about a "kind" of animal. A species is a population of animals that possess common characteristics and freely interbreed in nature and produce fertile offspring. Although species occasionally crossbreed, the hybrid offspring are usually sterile. The scientific name of a species may consist of two or three Latin words. The first word, the genus, is capitalized while the second word, the species, is not capitalized (Behler and King, 1979, 17).

"North American box turtles belong to the family Emydidae. This large family also includes aquatic and semi-aquatic turtles of the Americas, North Africa, and parts of Europe and Asia. American box turtles are separated from other turtles in this family by the genus classification *Terrapene*" (Cook, 2008, 10–11). There are four species of *Terrapene*: **T. carolina, T. ornata, T. nelsoni,** and **T. coahuila.** However, **T. nelsoni** and **T. coahuila** rarely become pets (Cook, 2008, 11).

The "**Terrapene carolina** has six subspecies including **T.c. carolina** (common or Eastern box turtle), **T.c. triunguis** (three-toed box turtle), **T.c. major** (Gulf Coast box turtle) and **T.c. bauri** (Florida box turtle). **T.c. mexicana** (Mexican box turtle) and **T.c. yucatana** (Yucatan box turtle) are rarely seen as pets" (Cook, 2008, 11).

"Within **Terrapene ornata** are subspecies **T.o. ornata** (ornate or Western box turtle) and **T.o. luteola** (desert box turtle). The two subspecies of **T. ornata**, along with the eastern, three-toed, Gulf Coast and Florida box turtles, are the American box turtles most commonly found in the pet trade. Each has a distinctive look, but ranges of several types overlap, and breeding between the species is not uncommon. Hybrid offspring are especially difficult to identify" (Cook, 2008, 11).

The Eastern box turtle "**Terrapene carolina**, seen in the pet trade, generally inhabits moist deciduous woodlands and other habitats close to trees and water. The forest floor is their home, and it provides them with food, nesting areas, and hibernation sites. During hot, dry summers, these turtles estivate or become inactive by hiding under leaf litter until moist conditions return. During the winter, they hibernate. (Florida box turtles may remain active all year.) All box turtles are diurnal and will retire to a safe place under tree roots, rocks, or low shrubs as evening approaches" (Cook, 2008, 11).

# Physical Features

"***T.c. triunguis***, the three-toed box turtle, is four to six inches (10.2 to 15.2 cm). It has a high-domed keeled carapace that is often more elongated and narrower than the Eastern box turtle" (Cook, 2008, 12). The shell color varies from olive, brown or yellowish brown and may have thin lines, dashes, and spots of yellow or brown. The plastron is typically a solid tan but it may have dark smears around the scute margins. The skin is brown with orange or yellow scales on the head, neck, and legs. The adult male has red or pinkish eyes. These turtles often have only three toes on the hind limbs, but four toes are not uncommon. They have a large range and are found in western Georgia and west-ward into Alabama, Louisiana, and northwest Texas and northward into Arkansas, Missouri, and parts of Kansas (Cook, 2008, 13).

The photo on the left (below) shows the underside or plastron of Charley.

The plastron is normally concave on all other male turtles and tortoises except the three-toed male box turtle, like Charley, which has a flat plastron. The plastron is flat on all female turtles and tortoises. In the photo on the right shows a *T.c. bauri* (Florida box turtle) male to show its concave plastron and inward curve.

# Gender

Sometimes, male and female box turtles are hard to identify by gender. There are specific traits that a male has, such as red eyes and a longer, wider tail. The bottom of the tail is the tail vent, also called the cloaca. The cloaca is the opening for the digestive, urine, and reproductive tracts. Further down from the marginal scutes. When a male shows his penis, it is one inch wide with purple-black coloring. Males would sometimes evert (turn outward) their penis while they are in the water or wet grass. This is "penis fanning," and they do it for reasons unknown, perhaps for cleaning or cooling. However, females have brown eyes, and the cloaca is closer to the marginal scutes and has a shorter tail (Box Turtle Site, 2019).

## Turtle Tails

**Charley**                                 **Celie**

# Eye Color

**Photo by Melissa Podlesney Maravell**

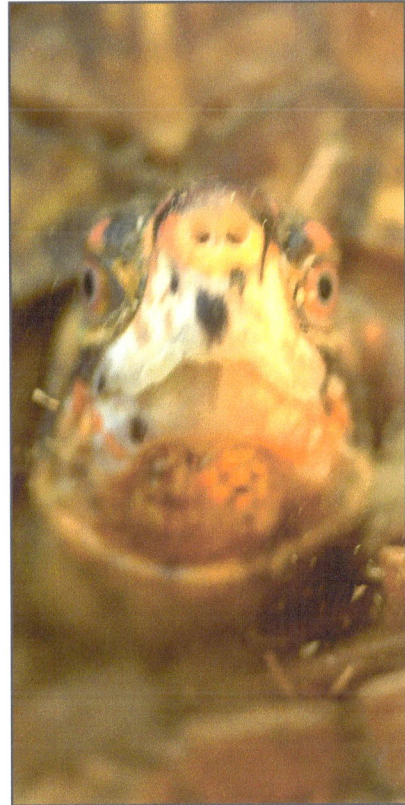

# Age

Dr. Jones informed me that box turtles can live 50 years or more. He told me in 2011 that since Charley was 31, he could easily live another 31 years. Charley is now 44 and is very active. Dr. Smith added that, without hibernation, box turtles can live well into their fifties. In the wild, they can live 50–70 years plus. I once took care of an 84-year-old box turtle on August 12, 2008, at Celeb Smith State Park, Smithtown, New York.

# Comparing Chelonian Species

I am going to compare box turtles, namely *Terrapene carolina carolina and* the red-eared slider *Trachemys scripta elegans.* The red-eared sliders are mainly aquatic turtles. The name "slider" is because they can dive into the water, and some slide off the rock edge. Most are active during the day and are often seen basking in great numbers when the sun is shining and the temperature is high enough (Patterson, 2004). Like all ectothermic reptilians, box turtles must regulate their core body temperature. Instead of burrowing like the boxer, sliders dive into the water to cool off. Sliders can stay underwater for long periods before coming up for air, unlike boxers who swim on top of the water and can easily be tired if they cannot find land to rest. Sliders require filters, water heaters, and heat lamps that should be over their basking areas. This area should be dry so that the turtles can dry off completely. Then, after their body temperature is regulated within the correct optimum range, they return to the water (Patterson, 2004). This way, they can keep themselves cool. Boxers burrow and sliders dive!

Enclosure requires a higher level for the turtles' basking area. I used flat rocks, not slates that chip off, for the turtles to bask on. I also used a water heater, a glass tube-like device, so it would not get broken under the rock and by the glass enclosure wall. Fish filters do not work because they are not constructed to do this job. Sliders can get messy, much more than fish. Sliders are known to keep their distance and are more easily frightened. They are better observed than handled, like box turtles, who do not like being handled much. Some personalities like Charley are very inquisitive, friendly, and outgoing.

Chinese box turtles *(Cuora flavomarginata)* are not related to Eastern box turtles but are similar to the Asian species. Both need a water tray in their enclosure. Asian species are semi-aquatic. They first came from the moist habitats of Asian rice paddies (Bartlett and Bartlett, 2001). These friendly turtles have a yellow margin on each side of their heads

My mom once pointed out how pretty they are: "That Boxy has a cute face." They truly are attractive turtles. Like Eastern turtles, they can close up completely, because they have a hinge on their plastron.

# Tortoises

There are two similar species, the red-footed tortoise (*Chelonoids carbonaria*) and the yellow-footed tortoise (*Chelonoidis denticulata*). The red-foots are medium size, whereas the yellow-foots are much larger. Both species were first imported from humid tropical climates such as South America. Then, after a while, they were bred in southern Florida due to the warm, humid conditions (Vosjoli, 1996). These tortoises cannot swim. No tortoise can swim. They are terrestrial (land) creatures. If you notice their back legs—don't they remind you of elephants' feet? "They have tree-like hind legs that end in oval feet with heavy cushioning at the bottom and short heavy claws" (Walls, 1996, 2). "If you see webs between the toes of the hind feet, you don't have a tortoise but instead a box turtle or some other water turtle that mostly lives on land" (Walls, 1996, 1–2). You can identify a tortoise by its size or color patterns. "Tortoises are not included in the Box Turtles genus *Terrapene.*" A tortoise is a "terrestrial turtle that lacks webbing between the toes of any of the feet and is a member of the family Testundinidae. With few exceptions, they are vegetarians in nature, feeding on variety of fruits, leaves, and even dry grasses. Only a few species are found near water, and even fewer regularly take significant amounts of meat in their diets" (Walls, 1996, 1–2).

They can also soak in shallow, warm water to hydrate themselves. However, in deeper water, they will drop to the bottom like a rock. Red-foot tortoises are friendly and personable, evolving characteristics. They can grow up to two feet long and are brightly colored, beautiful creatures. I have Eddy (Edwina), who has a yellow polka-dotted head, and Rosie, a cherry head tortoise. Both red-foots have hot orange-red scales on their legs and feet, giving them their name. I find them to be gentle, good-natured, and have sweet personalities.

# Chelonian Characteristics

"Turtles lack spaces between skull bones called fossae." They "have skulls that bulge out at the temporal region to provide attachment places for these muscles" (Flank, 1997, 11).

No turtle has teeth, but they have a hard, bony jaw sheath. I see my turtles use their front claws and strong jaws to tear off bite-sized chunks of food. I even see my turtles sometimes wipe off their faces. It is known that chelonians have excellent vision, can see color, and can detect motion. They also have a keen sense of smell. Chelonians use their eyesight as the primary method of finding food. As they are well able to see colors, they are particularly sensitive to reds and yellows. Mine is also attracted to the color orange. Since they are attracted to yams, they may also eat squash; they confuse these foods when these two foods are cooked in the same manner. Charley ate bites of the squash, then stopped and sniffed it twice.

Chelonians can also sense a range of infrared wavelengths that are invisible to man. That may explain why my turtles are sometimes active at night when they should be sleeping. I always get up to check on them with my searchlight whenever I hear an unusual sound. I also check their heat. They are kept awake with the red lights commonly used in heat lamps.

# Smells Sweet?

According to *The Turtle* book, the tongue of turtles is not the same as that of snakes and lizards, which extend their tongues out to detect airborne chemicals. Turtles, however, "are able to use their thick, fleshy tongues to capture scent particles in the air and transfer these to Jacobson's organ" (Flank, 1997, 12). This "organ is directly connected to the brain by the olfactory nerve....Turtles have a keen sense of smell, even underwater" (Flank, 1997, 19). The turtle can smell by "opening its mouth slightly, drawing in a small amount of water through the nostrils and passing this through the Jacobson's organ before expelling it from the mouth" (Flank, 1997, 12).

# My Shell-Ter, My Armor

A turtle travels around in its own house. This is similar to a trailer or RV. Their house is the shell; when threatened, they hide and close up for protection from predators. The turtle would "withdraw its head, neck, tail, and limbs completely within the shell. It does this by expelling air in the lungs to allow room for the limbs and by twisting its neck into an S shape to bring the head into the closing shell" (Dodd, 2002, 7). Box turtles form "a tight seal by closing the plastron (bottom shell) upward to fit snugly against the carapace (top shell)" (Dodd, 2002, 7). A movable hinge closes the plastron and carapace together (Dodd, 2002, 7). Turtles would stay closed up until they sense it is safe to come out. This bony armor is their shell-ter.

All chelonians have shells encased in bony armor, which is their protection and shell-ter. The shell consists of a carapace, the upper shell, and the plastron, the bottom shell. Bony bridges connect the carapace and plastron. "The carapace consists of some 50 bones derived from ribs, vertebrates, and dermal elements of the skin. The plastron evolved from the clavicles, interclavicles, and abdominal ribs" (Madar, 2005, 81).

"The bony shell is covered by a superficial layer of keratin shields called scutes" (Madar, 2005, 81). "Keratin is the same substance from human fingernails and hair are made. Scutes are made up of living tissue and contain nerve endings" (Flank, 1997, 14). If you touch a turtle shell, they can feel it (Flank, 1997, 14).

Scutes contain colors that vary in different species. "Both scutes and underlying bone are capable of regeneration. Turtles produce new scutes with each major growth period and retain or shed the scutes from the preceding growth period" (Madar, 2005, 81).

The turtle shell is attached to the turtle and cannot be removed. It "prevents the chest from expanding." Turtles have "a special set of muscles in the body to expand and contract the size of their chest cavity by moving some internal organs around, pumping air in and out of the lungs like bellows" (Flank, 1997, 13). When turtles breathe, their lungs move into the leg cavities. In my observation, I have seen their front legs move in and out a little when the turtles are relaxing.

# Slow Poke

According to Flank (1997), turtles are slow because of two major reasons: they have a three chambered heart consist[ing] of two atria and a ventricle, incompletely divided by muscular wall....Blood is pumped to the lungs by the upper chamber of the atria, returning to the lower ventricle, where it mixes with oxygen-depleted blood returning from the rest of the body. This mixture of oxygen-rich blood and oxygen-depleted blood is then pumped into the other atrium, where it enters the aortic arches and is distributed throughout the rest of the body. Because un-oxygenated blood returning from the body is mixed with oxygen-rich blood, or oxygenated blood, returning from the lungs, this causes turtles to tire easily, needing frequent stops to rest. (13–14)

The second reason is that Chelonians "have a characteristic walking pose, with their legs bent out at the elbows and knees, which makes it look as if they are halfway through a pushup. Turtles have their limb girdles inside their rib cage" (Flank, 1997, 17). This significantly limits leg mobility. Nevertheless, their legs are strong and can move quicker than you think!

# Manicuring

In captivity, occasionally, beaks need to be trimmed due to a diet of mostly soft food. A qualified vet usually files down the beak and carefully avoids going near the bloodline. Nails also need to be trimmed since there is no wear and tear on rocks, like in the wild. When trimming nails, always be advised to use Kwik powder if the nail starts to bleed. If you do not have Kwik powder, use a bar of moist plain ivory soap to form a clot or baking powder to stop the bleeding.

# Home, Home on the Range

Eastern box turtles (*Terrapene carolina carolina*) have a homing instinct. They can find their way back to their home range. However, this can be a problem if their former locality has been subjected to habitat destruction, especially if the woods have been cut down and that land has become a parking lot!

There are four ways reptiles use orientation:

1.  Visual clues or landmarks since most reptiles can see color.

2.  Using polarized light in orientation and navigation. It is the way light reflects on the soil. Reptiles can tell whether the soil is damp or dry by how the light reflects off it. This is especially important for humidity-loving reptiles and amphibians.

3.  The third way of homing reptiles is using chemical clues or olfactory nerves. They use Jacobson's organ to smell odors.

4.  Lastly, reptiles use magnetic orientation. We especially see this in sea turtles since there are no landmarks in the open oceans (Zug et al., 2001).

During my volunteer position at the Caleb Smith Conservation Park in Smithtown from 1995 to 1997, under one of their conservationists, I learned that to keep box turtles from wandering back to their home range, they must be hibernated (brumation), at least for one winter. (See chapter 8 for hibernation). It is very important that when seeing a turtle crossing a road, please place the turtle on the side of the road in the same direction that it was heading.

# Communication Common in Chelonians

Tactile communication is common in chelonians. This occurs when one chelonian rubs, presses, or hits a body part against another chelonian. Tortoises use head bobs, biting, ramming, and flipping over in communication (Zug et al., 2001).

"Tortoises and turtles use a combination of visual and chemical signals during social interaction" (Zug et al., 2001, 228). Visual displays involved with tortoises are head bobs, shows off patterns and colors on the forelimbs, neck, and head. When two tortoises meet, a male will perform head bobs or sway his head back and forth. If both are males, the other tortoise will respond with similar behavior. The interaction can escalate into butting, biting, or flipping over. Yet, these two tortoises will sleep in the same burrow at night and then start the biting episode again the next day. If one is a female, after the biting or ramming, the male will attempt to mount her, scratching her shell, grunting, and moving his head in and out of his shell. This behavior may or may not lead to copulation (Zug et al., 2001, 228–229).

The reader will discover Charley's life, which he had four near accidents and four life-threatening illnesses but was spared time after time.
If Charley could talk,
this would be his Turtle Testimony.

# Chapter 2

# Short Story Section

HYLAND

# Now meet Charley, my cherished chelonian!

Charley's so cute in his orange suit!!!

# The Happy Herp Home Coming!

**October 18,1985**

My mom knew I was making many phone calls to different pet stores to find out which one had turtles. I finally found one such store in Medford. My mom said, "Let's go get a turtle this week!"

Friday, October 18, 1985, was a beautifully sunny fall afternoon. My mom, dad, and I got in our red '66 Caprice and drove to the store. When we arrived at the store, we looked at the sign that said, "Country Critters."

My dad said, "There's a Critter waiting for you here!"

As we walked into the store, my mother saw the tank with the turtles in it. Standing on some rocks, one turtle walked right over to my mom, looking at her through the glass. Mom said, "This one's cute."

I approached the tank and saw the cutest orange face staring at me. The turtle stretched his neck out, held his head high, and pressed against the glass to see me. I have never seen a more beautiful turtle in my life. He had various shades of orange on his head and neck.

I asked the salesperson if I could hold that turtle and the tortoise in the same tank. I walked over to my dad, holding the two chelonians. The tortoise never stuck out his head, but the little box turtle was very alert. His head stuck out, looking straight at my father.

Dad asked, "Are you going to get him?"

I said, "I'm not sure."

He replied, "What do you mean, you're not going to get him? He's so friendly."

So, I said, "I'll get him."

That day, we paid $23 for Charley and some supplies. They carefully placed him in a box, and we drove home. I am so thankful I took my dad's advice that day. It was one of the best choices I have ever made. The day I brought home my buddy Charley was a wonderful day indeed!

The store told me that Charley was five years old and had come from Florida. He would live many years with me, Mom, and Dad. Years later, I found that store receipt for $23. Then, I would spend that and much more on vet bills. But it was worth it all!

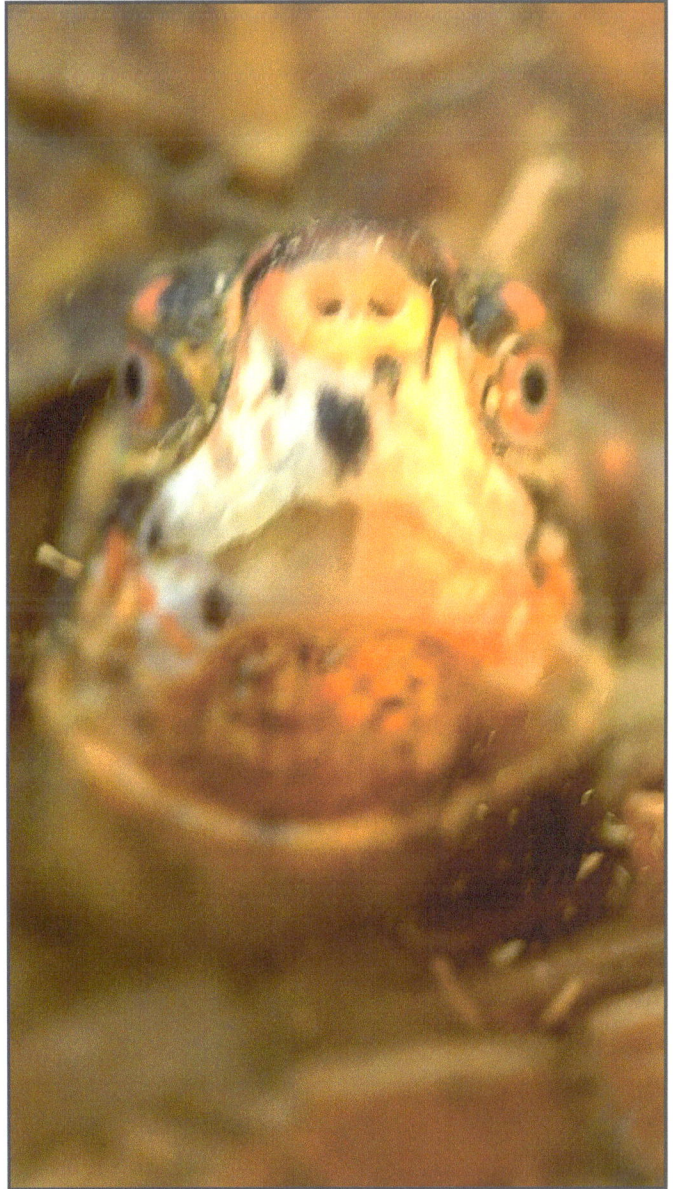

# First Trip to the Herp Hospital

When I first got Charley, I thought he was a female because of his small tail. So, I named him Charlina, after another box turtle given to me in 1967. I had her for 18 years and remember that she had yellow on her neck and sweet brown eyes.

One day, I thought Charley's large intestines were coming through his tail. So, I searched for a herp hospital with a veterinarian who treated turtles. I found one and had my dad drive me there. The vet showed me his herp photo book and talked to me for a while about Charley. He then asked me if I had seen any blood on Charley, and I said, "No."

Then he showed me another picture and asked me, "Did it look like this?"

I replied, "Yes."

The vet smiled and told me Charley was a male turtle, and it wasn't his intestines but his penis from his cloacal or tail vent.

# Close Call, 1987

One summer day, my mom told me to check on Charley, so I came out and picked him up. It was so hot that I saw my boy Charley's mouth and nose had bubbles. I quickly grabbed the garden hose and completely revived him. Thanks to my mom, Dorothy, who saved Charley's life that day! Chelonians can bake in their shells and suffer from heat stroke if left out in the blazing sun. After that incident, I became very protective of Charley and hoped not to leave him out there again. That was too close a call for both of us! (Concerning close calls with thermoregulation and POTZ—Preferred Optimal Temperature Zone— see chapter 4).

# When Charley Met Peggy in 1988

Charley was basking in the green grass of an enclosure, enjoying his time outside. Then suddenly, Peggy, my dog, jumped into his cage and put him in her mouth. Then Peggy dropped Charley on the top step near my door. My mom opened the door, saw what she thought was a rock, and wondered who would put a rock here. As she was going to fling it, she saw little Charley's back feet kicking.

"Jeanne! Jeanne, it's Charley," she called. "Come here and see what I found on the top step."

I came running screaming, "Charley! Charley!" Next, I took my orange boy to the herp hospital.

A vet examined him and said, "He's alive. It takes strength to hold himself in like that. You have one very scared turtle on your hands."

When I got home, I placed Charley on my lap and gently said his name, "Charley? Charley, it's all right."

Charley did not come out of his shell for at least an hour. When he heard my voice for a while, he finally sensed it was safe to come out of his shell, and I held him in my hands.

# In Becky's Big Mouth! Summer 2002

Charley was soaking in what I called his Summer Reptile Resort. While I was nearby, my other dog, Becky, grabbed him in her big mouth. Oh boy, here we go again! I was so scared for my little boy that I screamed and pulled him out of her mouth. It was another close call.

# Escaped Explosion—June 22, 2004

A blue heat bulb from Charley's tank turned on. Then the heat bulb exploded, and glass flew everywhere. Thankfully, I was right there. I grabbed Charley and checked him over carefully. I knew my boy was probably scared, so I held Charley for about half an hour, calming him. Then placed him in a dry soak pan, in his little bed. I spent seven hours cleaning the tank to ensure all the glass was out. I even asked my mom Dorothy to check the tank before putting Charley back in there. I continued cleaning until 1:30 a.m. to make sure every tiny little speck of glass was out of his herp home.

While cleaning, I noticed Charley was sleeping for hours, poor little guy. Finally, I felt I had gotten all the glass out of the tank and put an all-new green reptile rug in there. Then, I put the bed in and placed Charley in a clean herp home. He fell asleep quickly, knowing he was safe when I was there. I never used those heat bulbs again and started using ceramic heat emitters, which are safe when misting the tank.

Jeanne Hyland

# Tripod Triunguis—April 2, 2010

I noticed Charley's right back leg was in his shell. I put him in some water, but it still did not come out. So little Charley walked around on three legs for two and a half days. He even managed to climb in and out of his reptile pool on the ramp. I thought he may have injured his right back leg. I called Dr. Jones's office and spoke to a technician, who told me the doctor was not in the office. The technician felt it was nothing to be concerned about as long as Charley was eating properly and was active. He also said that Charley's leg would come out of its shell in a few days.

However, his leg still did not come out. I made an appointment with Dr. Jones's office. After the vet examined Charley, he said he was concerned about this leg. If his leg didn't pull out soon, he would have to give him pain pills. If that doesn't work, he'd do X-rays. Then, Dr. Jones used his tools and carefully tried to pull the leg out. He even asked me to call for another assistant.

As he continued to pull gently on Charley's leg, a pile of eco-earth fell onto the examining table. Growing two inches high and two inches long.

Then he took Charley and placed him in a soaking pan. I asked the assistant to please fill it up with water. Soon enough, Charley was kicking both his legs.

With a dry sense of humor, Dr. Jones said this was the first time he'd ever done an eco-earth impaction. He told me Charley was fine now, and I could take him home.

Now when Charley gets his daily soak every morning, I make sure to get all the eco-earth out of his leg cavities. We have had no further problems with his legs.

# Speedy Slider—June 2002

One June afternoon, I decided to put Eddy, the red-eared slider, and Charley into a beautiful blue pool, nicknamed the Reptile Resort. But I first placed the turtles in a small pan. Then, as I turned away, Eddy saw his chance to roam freely. He quickly climbed out of the pan and scooted across the cement patio. Then walked even faster across the green grass in my backyard until he disappeared. Where did Eddy go? Would I ever see him again? I was so frightened for Eddy that Mom and I desperately searched for Eddy for two long hours. Finally, I took the rag that Charley slept with and held it in front of Becky, the dog. She sniffed the rag, and I told her, "Go find the turtle."

Becky started to sniff all over the backyard, then scratched behind a big bush. I looked behind it, and there was Eddy! Imagine that Becky found my Eddy in less than five minutes!

I picked him up and examined him. He was fine.

# Chapter 3
## Reptile Resort

HYLAND

# Charley Meets Eddy!

On the right is Eddy: Red-Eared Slider (*Trachemys scripta elegans*) and Charley.

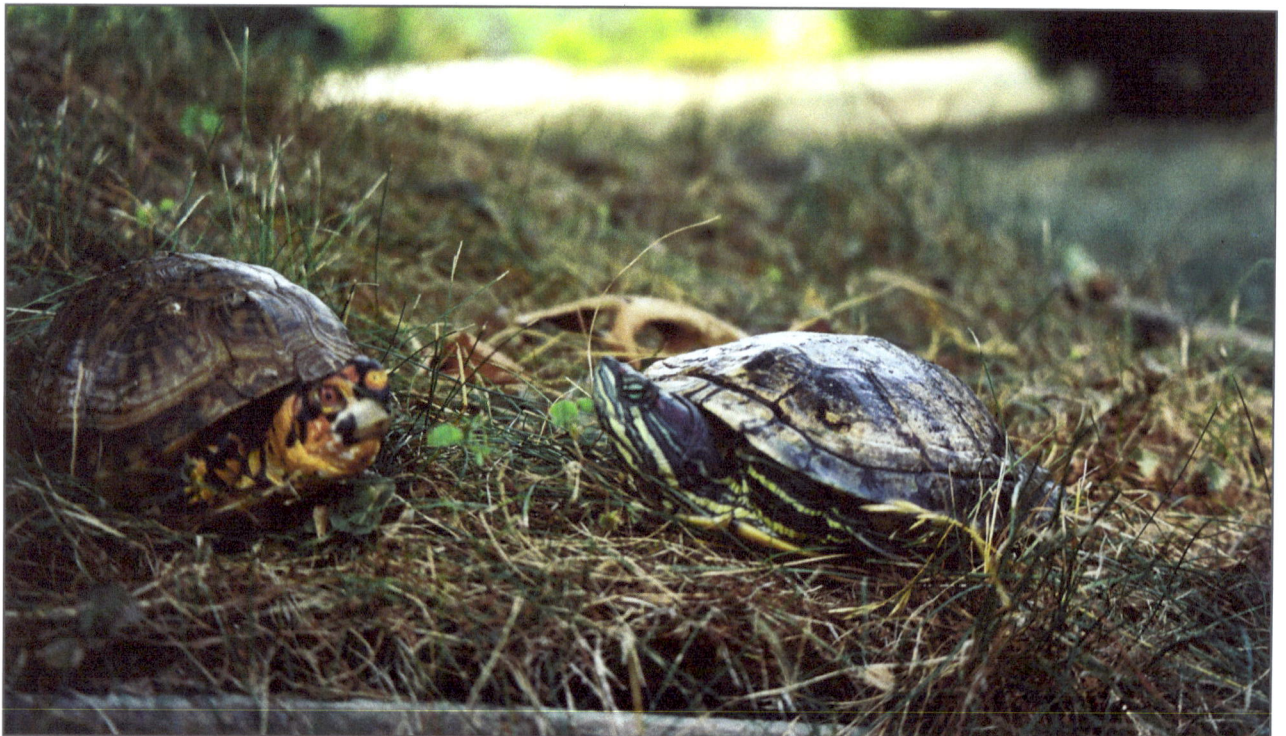

Jeanne Hyland

# Reptile Resort—Summer, 2002

## A fun day in the sun with Charley and Eddy!

Jeanne Hyland

# Chapter 4
# Herp Hospital

HYLAND

# The Terrible Time for a Turtle

## Around February 2003

My chelonian Charley developed bubbles from his little orange nose. Due to the snowstorm, I wasn't able to leave the house. So I contacted the herp hospital, and the vet came over. He administered a needle in Charley's front leg with Baytril (an antibiotic). The vet informed me that Charley had a mild upper-respiratory infection. Below is an X-ray of Charley when he had an upper-respiratory problem.

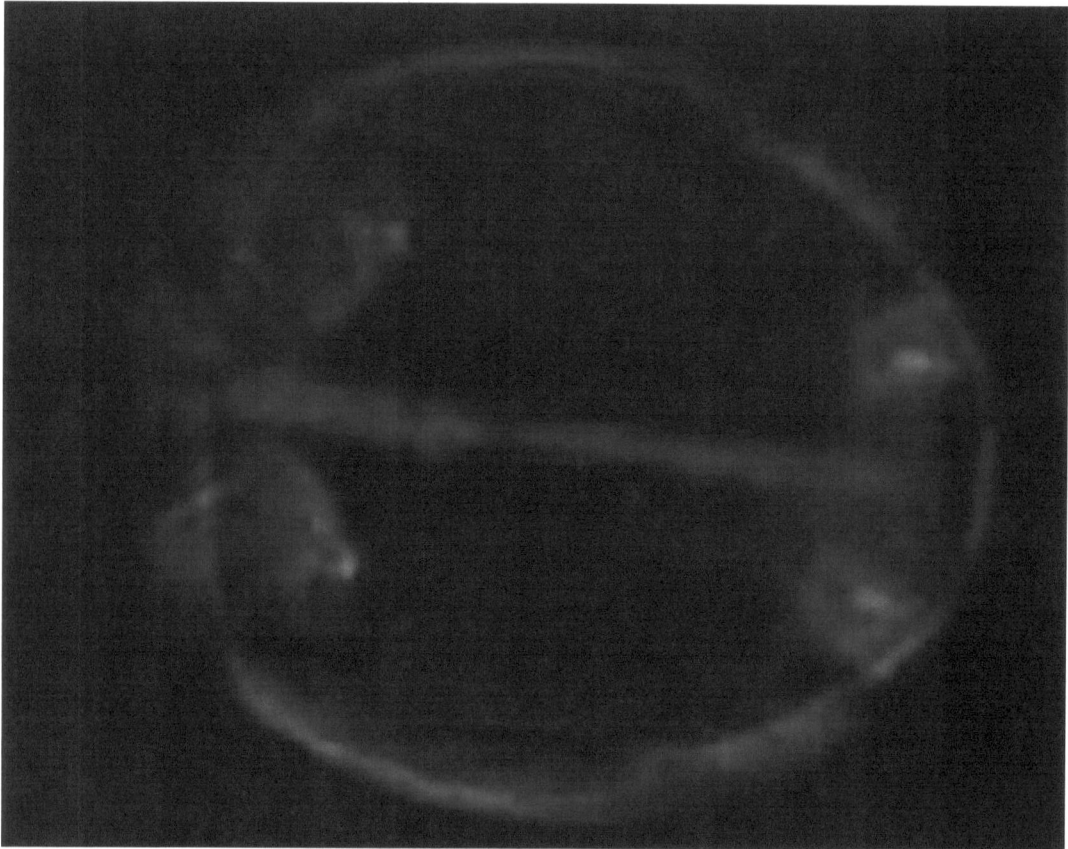

## Charley's Pneumonia

In April 2003, Charley was introduced to a Chinese box turtle named Boxy. By June 2003, I noticed that Charley had little bubbles coming from his nose, a sign of respiratory problems. The vet informed me to watch his diet and take him out in the sun. However, the bubbles became constant and didn't clear up. My chelonian caught pneumonia from the lack of proper quarantine when obtaining Boxy the turtle. I was wrongly informed about quarantine time methods and thought back then that a clinical check-up for the turtle was sufficient. Now I know that a three to six months quarantine is a much more effective way of protecting an already-established collection of herps. He had to get an antibiotic injected in alternating front legs for about three weeks. Never inject in the back legs, which would go right to the liver. During that time of antibiotics, Charley didn't have any appetite. I would withhold water for a few hours and make him drink the dandelion pulp. Eventually, I never saw any bubbles from my orange boy Charley.

On September 29, 2003, I took Charley to the herp hospital. They performed a fecal analysis and flotation. Charley had no nasal discharge at that time. It is not uncommon for captive turtles to become ill with pneumonia. Pneumonia, also called respiratory illness, can lead to life-threatening complications if not treated. The causes of pneumonia are deficiencies in husbandry, sanitation, and nutrients. A clinical diagnosis should be made based on the sound principles of clinical medicine, history, physical examination, and various ancillary diagnostic techniques (Madar, 2005, 865).

## Temperature and Immune System

"One mechanism for combating bacterial infection is elevation of body temperature, because high temperature inactivates or kills bacteria. Some lizards, snakes and turtles behaviorally select and maintain body temperatures significantly above their normal activity temperature. This behavioral fever mechanism appears to reduce infection and improve the reptile's resistance" (Zug et al., 2001, 296). No reptiles can produce a fever.

The preferred optimal temperature zone (POTZ) is the range from the high range to the lower range in temperature. So the maintenance of the (POTZ) "is critically important for normal respiratory function and normal immune system activity" (Madar, 2005, 865). According to the authors, thermoregulation in ectothermic vertebrates, "under conditions of normal activity, Amphibians and Reptile cease activity when they cannot maintain body temperatures within a specific range. The activity range is bounded by the voluntary maximum and voluntary minimum temperature" (Zug et al., 2001, 178). It is significant "when on medication turtles should be housed at the high end of the P.O.T.Z. to ensure proper drugs uptake by the body" such as antibiotics, etc. (Cook, 2008, 38). Hypothermia is too cold, which can cause respiratory problems and pneumonia. Also, hyperthermia is too hot, which can cause lethal levels and death.

Humidity is often overlooked as another factor causing pneumonia in reptiles. Certain species require more or less humidity than others. When researching what each particular species needs, it is important to know its scientific and common names to identify them correctly. Also, a nutritional imbalance with hypovitaminosis A (see chapter 7) is often associated with respiratory diseases in reptiles. "Inadequate dietary vitamin A results in squamous metaplasia of respiratory (or epithelial cells and the ducts of the mucus membranes) glands. Such changes may either predispose the respiratory tract to opportunistic bacteria or cause clinical signs that mimic pneumonia" (Madar, 2005, 865). Many pathogens that cause pneumonia are opportunistic because of poor hygiene. These pathogens eventually wear down the immune system of the reptile (Madar, 2005, 865).

# Stethoscope Use in Chelonians

Through physical examination "nasal discharge may or may not be associated with pneumonia. An upper respiratory component is not unusual in lower respiratory disease. Therefore, the presence of nasal discharge is not a sign of pneumonia but a reflection of disease within the upper airway oral cavity" (Madar, 2005, 866).

The best method of using a stethoscope on a chelonian is by using "a moistened gauze sponge over the diaphragm of the stethoscope." This facilitates auscultation of the lungs. "The clinician should pay particular attention to the presence of abnormal sounds and asymmetry during the auscultation" (Madar, 2005, 866–867). In chelonians, this damp gauze is placed on top of the carapace over the lungs. This helps the vet hear through the animal's shell. It should be done before turning to X-rays to determine the problem. A pediatric stethoscope should be used, as it blocks out the other sounds to hear breathing through the carapace (Madar, 2005, 866–867).

# Treatment

A form of treatment that can help with pneumonia is supplemental oxygen therapy. However, it is not always advisable in all pneumonia cases. One of the driving parameters for reptilian respiration is P02. Significantly elevated environmental oxygen invasions may inhibit respiration, thereby compromising the reptile, which is already having difficulty eliminating inflammatory debris from the lower respiratory tract. If oxygen therapy is to be provided, 30% to 40% oxygen levels should be sufficient (Mader, 2005, 876).

Supplemental oxygen therapy should be humidified during administration and ministration to avoid irritating respiratory mucus membranes, which could "compromise already marginal respiratory abilities" (Madar, 2005, 876). If the underlining causes of pneumonia are not corrected, such as utilizing proper husbandry, all the therapy given could be useless.

# A Glorious Day in the Sun

**August 23, 2003**

I took my boy Charley out to bask in the glorious sunshine. I basked him in the direct sunlight for eight hours, except when I placed him in the shade to cool off. I made sure Charley had enough water to keep him hydrated. While carefully making sure he metabolized all his calcium from the injectable, I later discovered that this procedure was not proven. It is unnecessary and can be fatal, as injectable calcium has a narrow range of safety.

Later, I was told by two vets who treated Charley that it was no longer necessary to treat him with these supplements as long as his diet was adequate. Adult turtles are slow in growth; full-grown turtles don't need calcium supplements unless they are lacking it in their diets. (See chapter 7 for calcium injectable.)

# Microflora

Microflora is the gastrointestinal healthy bacteria used for digestion that is responsible for the hind gut. According to *Reptile Medicine and Surgery*, "unlike mammals, reptiles must bask in radiant heat to maintain a core temperature to maximize fermentation in the hind gut" (Madar, 2005, 150). In other words, reptiles have to be warm enough to digest their food.

# Endotherms and Ectotherms

Endotherms such as birds and mammals use about 98% of the energy from ingested food. This energy goes into temperature regulation and activity. Their bodies produce heat for proper metabolism and energy (Zug et al., 2001, 191). However, ectotherms are reptiles and amphibians. These creatures "do not use the energy produced during metabolism to maintain body temperature. Their body temperatures are low for at least portions of the day and season," so "they have a low energetic cost of maintenance. Approximately 40% to 80% of energy ingested in food goes into the body tissues of ectotherms" (Zug et al., 2001, 191).

"The high densities and biomass that amphibians and reptiles achieve, even in low-resource environments, can be attributed to this" (Zug et al., 2001, 191). I believe that is why reptiles have such longevity and grow to such massive sizes. It may be due to their slow metabolism.

# Basking Thermoregulation

Reptiles are ectothermic; the term is derived from the Latin words for "outside heat," which better describes how all metabolism functions in reptiles. Their metabolism is the biological process controlled by a class of enzymes that work best in higher temperatures (Flank, 1997, 7).

"Ectotherms cannot produce enough body heat metabolically to maintain their core body temperature." All reptiles need an external heat source to keep their internal core body temperature high enough. "This is why turtles are most often seen basking on logs or rocks," taking in the natural "heat provided by the sun to raise their body temperature to an acceptable level" (Flank, 1997, 8).

# Shell Rot

In a few months, he had trouble with his shell, and his scutes were peeling off, leaving white patches of bone showing (shell rot). Infections of the shell can cause bacteria and fungus to grow. If the bacteria is wet rot, it is called ulceration disease; if it is dry, the rot is fungal. These infections can combine. Shell rot needs immediate attention as it can lead to deadly infections (Ebenhack, 2012, 69). Therefore, for ten months, I would rub a cotton swap of iodine over Charley every evening before I put the lights out. Then in the morning, before his daily soak, I would completely rinse off the iodine. By October 2003, Charley's shell had completely healed.

**"My Shell had Boo Boos."**

# The Infected Beak

On October 3, 2003, Charley was admitted to the reptile hospital in New York. I felt very discouraged until I found a card with a turtle on it that said, "Keep looking up!" in the bookstore. When I returned to the vet's office, I spoke to the vet, who said Charley would have a full recovery. Also, "it was not a fungus but a beak infection. The oral Baytril (antibiotic) is working well, and Charley should have a long life."

Charley was still hospitalized for two weeks, so I asked the vet, "What caused the infection?" He felt it was from something in his intestines. I visited Charley every day except Sunday. One particular day, I noticed a grey feather fell into his cage from the cage above in the exotic room. I laughed and joked, "Oh, Charley, you are growing feathers now!" I held him and talked to him.

On Thanksgiving Day, Charley's beak was still infected, and the oral Baytril (antibiotic) was not effective. Later, I learned about the infected beak and how it needed to be cured.

# Ride to Recovery

On December 3, 2003, I drove into Nassau County to the veterinary hospital in New York before the first winter freeze, where Charley would stay for four weeks. On the first day of examination, Dr. Charos said, "This infected beak has to be filed down so that the live beak can grow out." Charley had to have his beak filed up to the bloodline, but no further than that. He was also given assisted feeding through a tube to avoid side effects from the antibiotics, namely Amikacin. Amikacin is an antibiotic used to treat bacterial infections.

# Charley's Medical Records 2003–2004

Herp Hospital, NY.

Rehabilitated from:

Lower respiratory—Pneumonia

Upper respiratory problem

Shell rot: Septicemic Cutaneous Ulcerative Disease

**Monday, August 11, 2003**

"Charley's blood tests are normal, thank God. His liver and kidneys are normal!

Sept.29 03 Stool samples are negative.

October 6, 2003 (Medical) Oral Baytril 1 cc was given every day for seven days.

October 17. Charley was discharge from Herp Hospital, New York

Accession No. NY27886081

Owner: Hyland; Pet Name: Charley; Received:12/04/2003

Age: 18Y; Reported: 12/06/2003 03:58 p.m.

Test Requested—Results:

   Sensitivities

Amikacin: yes

Culture & Sensitivities: Source, Cloaca

Organism #1: Enterobacter cloacae, Heavy growth

Organism #2: Klebsiella pneumoniae, Light growth

Organism #3: Gamma (non-hemolytic) Streptococcus species

Enterococcus (non-pathogenic flora), Heavy growth

\*     \*     \*

Charley's beak infection needed to be treated as soon as possible because it was getting worse daily. Records show on December 5, 2003, that the turtle is doing well, and his necrotic beak was to be reshaped on December 6. All the infected parts of the beak were filed down, right before the bloodline.

Amikacin .025 given IM (intermuscular) on December 6, 2003.

To avoid abscesses, the Amikacin was given every other day in one front leg and the next time in the other front leg, but never in the back legs. Whenever he was not eating on his own to avoid the side effects of the Amikacin, he was tube fed four cc mash bi-daily, BID.

The term "force-feeding" has been replaced by a new term "assist-feeding." This new term implies a more sophisticated, calculated, and gentler approach to feeding the animal (Madar, 2005, 535).

Assist-feeding may be needed in hospitalized reptile patients for nutritional support. Feeding tubes, used with ball-tipped stainless-steel feeding needles, facilitate assist-feeding. "Involuntary feeding is managed by assisting feeding with indwelling tubes. It should be attempted only after dehydration and electrolyte unbalance have been corrected" (Madar, 2005, 277).

"Assist feeding is accomplished by gently opening the patient's mouth with a speculum, laboratory spatula, credit card, or radiographic film. A speculum may be used to hold the month open and care is taken to avoid injuring teeth or keratin surface in (toothless) chelonians. Food is passed as boluses or more commonly, as slurries administered via syringe, metal feeding tube or catheter of plastic or red rubber. Some patients may be lightly sedated for this procedure. For tortoises and turtles, the bolus may be gently pushed down the tract with a blunt instrument. The author prefers less-traumatic slurries to bolus feeding" (Madar, 2005, 277).

Charley was hydrated with warm water soaks and kept in a heated cage. (This is known as behavior fever to thermally regulate because reptiles do not have sweat glands. Heat causes the antibiotics to work.) During his entire stay of four weeks, he only had droppings in the cage on two occasions because he is "housebroken" and mostly passes stool in the water.

**December 8, 2003:**

Records indicate that he is doing okay with assist-feeding. As the days go by, his weight goes up.

**December 26, 2003:**

The records state "doing well." Reculture tonight. Owner requests he be given his favorite food, which was box turtle can food.

**December 29, 2003, 11:15 a.m.**

Wt. 290 grams

Eating, drinking: fair.

Amikacin: .025 cc given IM.

**December 30, 2003, 7:20 a.m.**

Wt. 310 grams: gained weight. Seems OK.

**January 3, 2004, 10 a.m.**

Urine—yes, Stool—no

Hydration, OK. Wt. 300 grams

Warm water bath BID in heated cage.

**January 6, 2004:**

Wt. 300 grams

All is well, and Charley is discharged from the vet's office.

*Avoid Ivermectin* which "can cause depression, paralysis, coma, and death in chelonians" (Madar, 2005, 1068).

# More Information

Signs of dehydration: A vet technician at an animal clinic counseled me that if a turtle's urine is yellow-green, or green, it is a sign of dehydration. If the urine is yellow, it is better; if it is white and clear, it's perfect. I always hydrate my animals with a daily soak in lukewarm water. This is also true with inadequate humidity or inappropriate water provision.

# Correcting Dehydration

If a turtle is too weak and reluctant to soak or drink, orally administer Gatorade or Pedialyte, as "these liquids offer easily absorbed fluids, electrolytes, amino acids, and sugars. Administering the fluids with a ball-tipped feeding needle attached to a syringe is preferred. Hydration needles should be corrected before addressing nutritional requirements. See a vet if the turtle doesn't drink on its own after forty-eight hours (Vosjoli & Klingenberg, 1995, 62–63).

# Charley Cured

On January 6, 2004, Charley was able to come home just before the next winter freeze. On the way back home, the heater in the car was not working, and a turtle needs to be kept warm! So the veterinary associate hospital staff wrapped him up in a towel next to a surgical glove filled with hot water. Charley was all bundled up for that long two-hour trip back home. I thanked Dr. John Charos and the staff for taking such excellent care of my orange boy, affectionately referred to by Dr. Charos as the "kid" for those four weeks.

I bought Charley a new home. It was a 40-gallon tank instead of a 20-gallon tank. The following March, I took Charley to see Dr. Jones. He took Charley's bloodwork to see if he had too much Amikacin. Although the infection was gone, the results proved it was perfect, proof that he was finally cured.

Just a reminder: Sometimes, overgrown beaks that chip but are not infected need to be filed down. For example, my box turtle Celie had her beak chipped and was filed down by Exotic Animal Hospital in Orlando.

Happy Herp!!

# Chapter 5
# Green Serene

HYLAND

# The Homecoming

When I arrived home from the hospital with Charley, I first let Charley bask for 20 minutes under a heat lamp. Then I brought him over to his new 40-gallon breeder tank; I placed Charley into his new, large herp bed, which I called "Charley's Cozy Cot Corner." It had a pillow with a pillowcase covering it with a lot of room to roam. It was all covered by a soft and comfortable green terry cloth towel. During the day, my little boy Charley would walk around in his tank, and then, at night, I would set up his bed. Charley was living his best life!

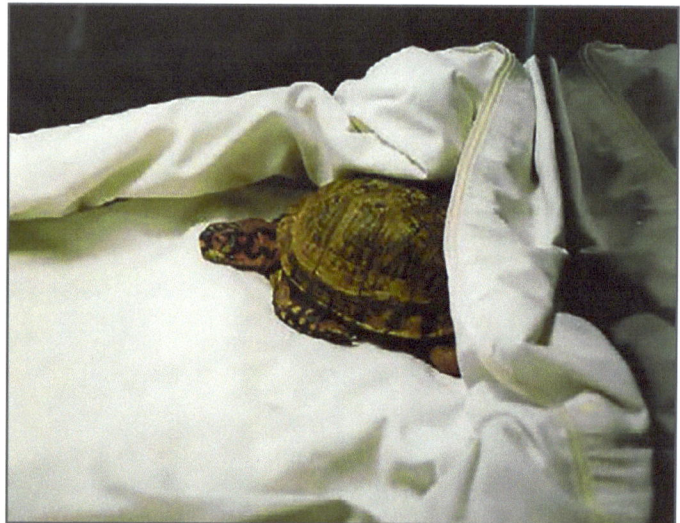

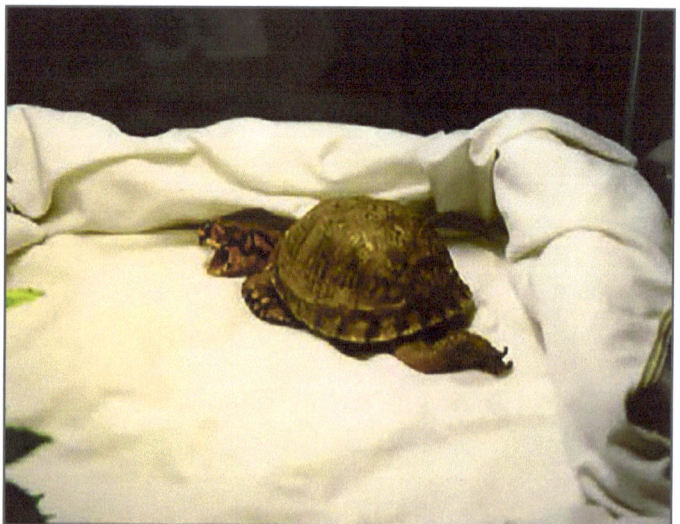

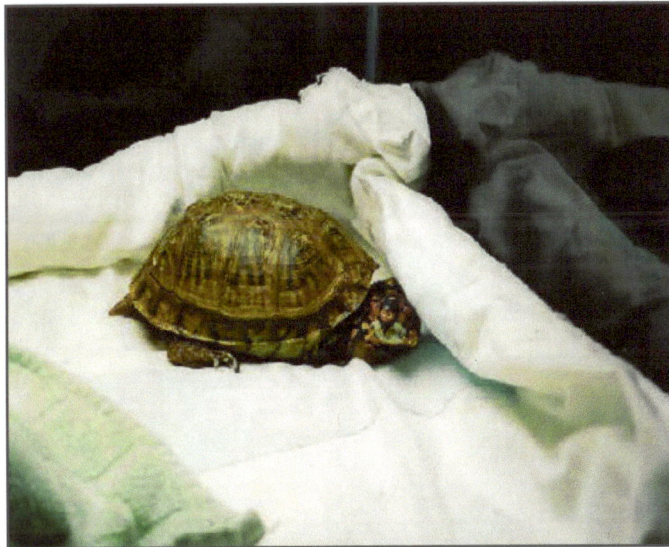

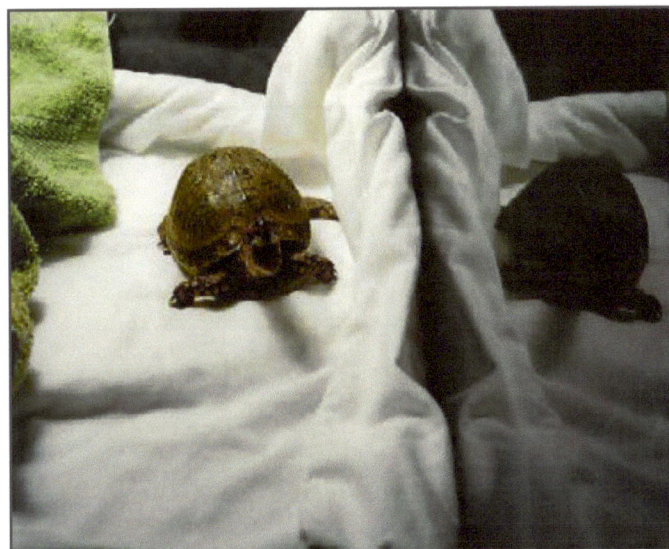
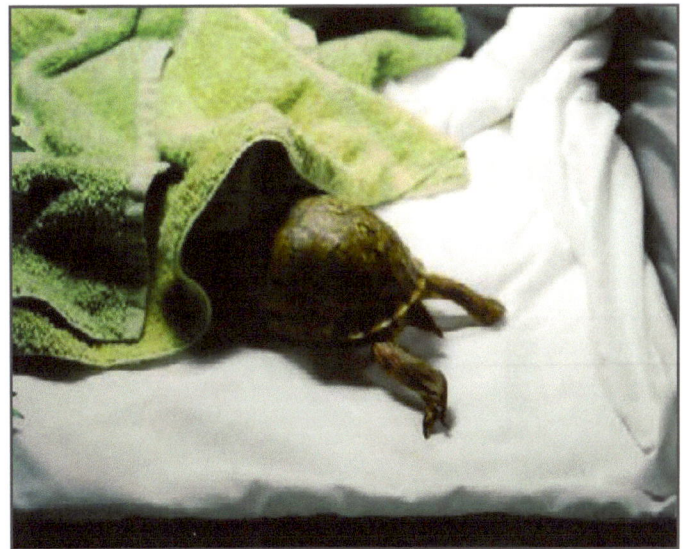

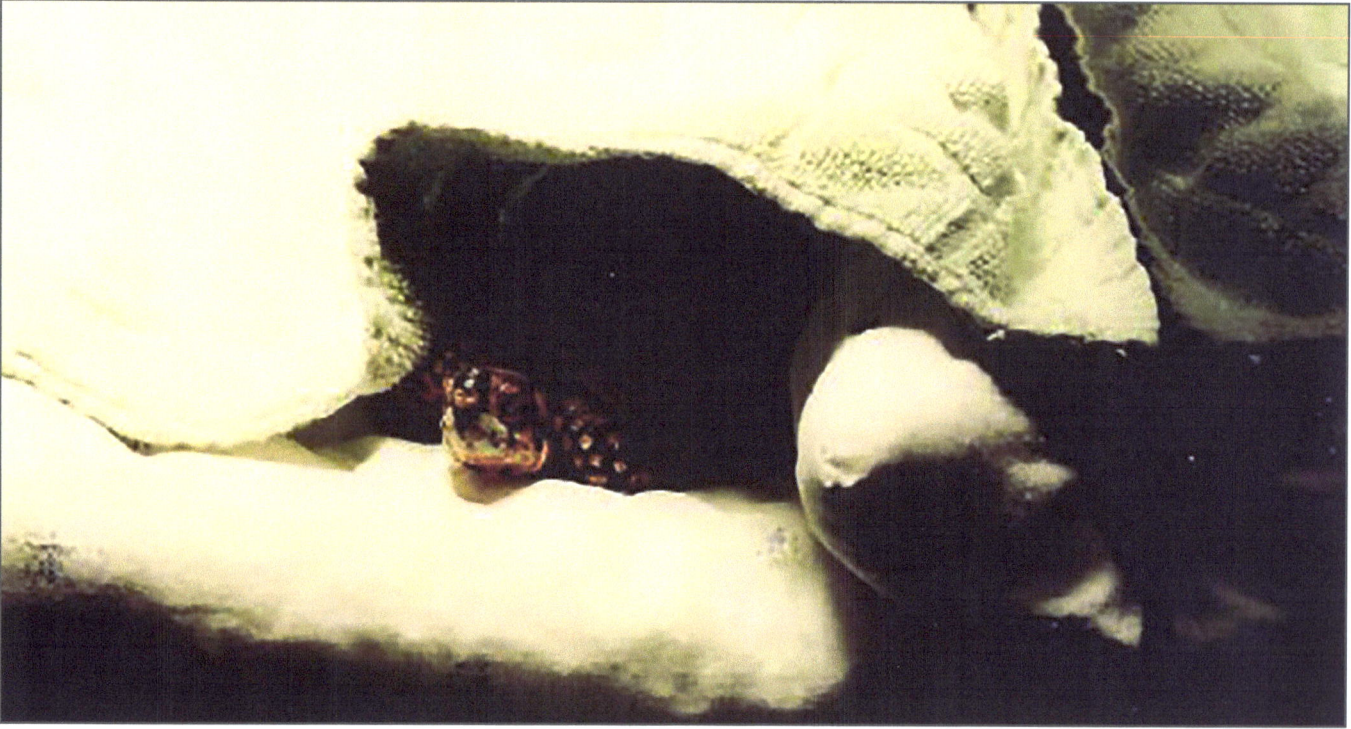

# Nighty-night time!

# Chapter 6

## Herp Hatch Day
### October 18, 2005

HyLAND

On that happy day, I celebrated Charley's 25th birthday with my mom. Unfortunately, my mother Dorothy left home for the hospital the next month. Ironically, she died in June the following year, near the same area where Charley recovered. Celebrating Charley's birthday that year was very special, and I took pictures and videos of him with my mom. No one knows the future; we must enjoy what we have when we have it!

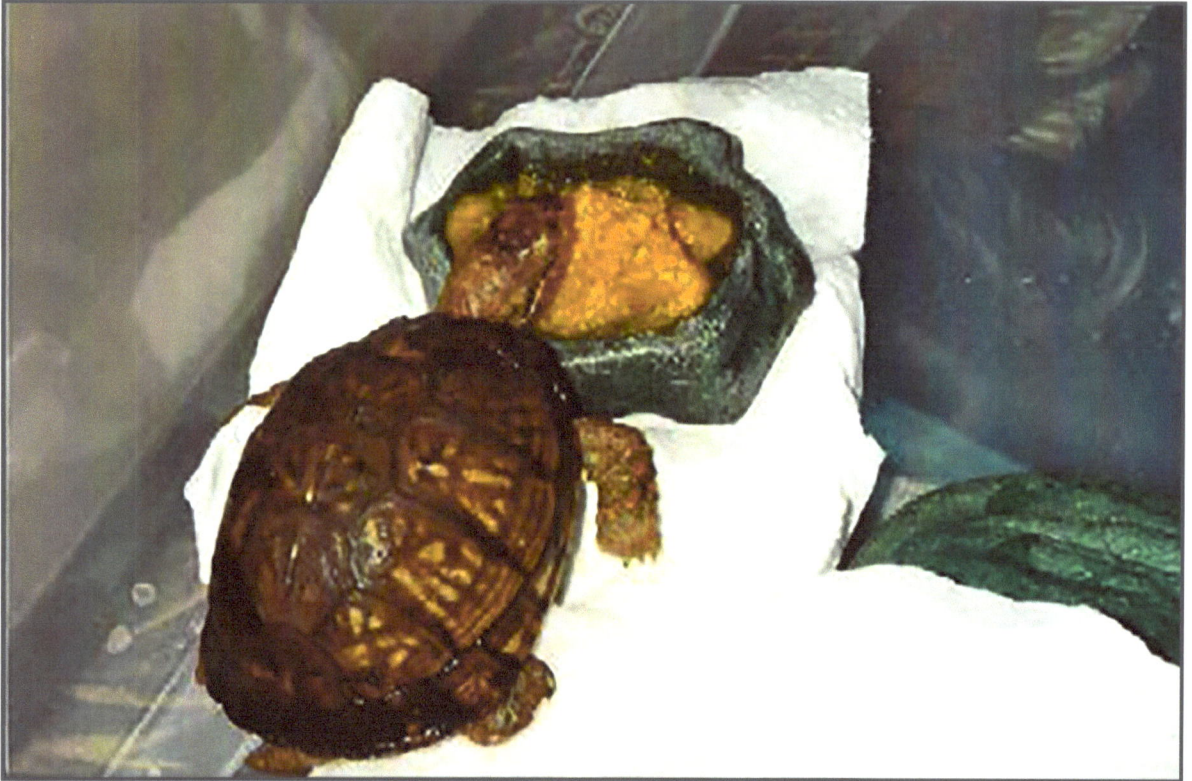

**Centereach Long Island, New York**

# Chapter 7

# Nutrition

HYLAND

Nutrition is the one factor that needs to be adjusted from time to time. Nutritional needs don't change with age. Over the years, I changed Charley's food in the 1980s. I was unaware of all their nutritional needs, except for earthworms. I was told that common dog food, Alpo, and lots of bananas, was an acceptable diet for a box turtle. My understanding of the animals' nutritional needs has come a long way since then! Also, herpetology has significantly improved the scientific research on herps' nutritional needs.

Box turtles are often finicky eaters in captivity and very challenging to feed. This is because when in the wild, they would take advantage of seasonal plants and insects until they were gone, tired, or bored. I have been aware of Charley's finicky eating habits for over 38 years now and have learned his preferences.

How to fool fussy, finicky eaters? I would mix a variety of foodstuffs together. Variety is essential because they soon learn to turn down foods that comprise a balanced diet. I would rotate the food selections to keep a good variety of meals. For example, some days, I have made up to three meals for Charley. He turned down the first two meals and only ate the third meal. That's finicky! But still, even though he is a fussy eater, I sure do love my Charley!

Another time, I showed Charley two fresh, lively earthworms. I placed them on his small feeder dish, and he showed no interest in them. I wiggled them and showed them to him for quite a while, but he did not want them. So, I put them back into the container and proceeded to bring him two lively gut-loaded crickets. Nope, he did not want them either!

After a while, I rinsed off a plum and sliced it nice and thin. I put the plum on his feeding dish, picked the pieces up, and dangled them near his face. No dice! He did not want this either! Finally, rinsing off his utensils, I prepared and soaked Mazuri tortoise pellets, then placed them in his feeding dish. Charley decided that he'd have a drink of water before eating.

I started calling him, "Charley, oh Charley, come and eat!" I'll never forget how he held up his head so straight and high, very alert and came back over to his food dish when I called. Would he eat? Ah, yes! This was what he was waiting for.

On another day, I held the crickets for Charley. He decided to eat one lively adult gut-loaded cricket. Surprise, surprise! I never knew what mood he was in. Unfortunately, I could not ask him what he would like to eat today! Crickets are fast, so I held them for him. Later, you will learn about gut loading.

"Box turtles are not fond of vegetables," so wash the items and chop all items into bite-sized pieces (Madar, 2005, 90). I tried to mix vegetable greens in a moist mash—pellets soaked in water—but my box turtles knew to pull out the pellets and leave the greens. Finicky eaters for sure!

Anyone who does not have patience and time should not start out with a box turtle. Box turtles are very finicky and will eat only the food they prefer that day. A box turtle owner needs a variety of food, lots of time, and a lot of patience to have a happy, healthy herp!

# Animal Protein

Protein is very significant in turtle growth, especially for breeding females. However, protein should be given in moderation. It is recommended that red-foots eat their portions of protein once a month. Too much protein can contribute to pyramiding. Also, a lack of protein can cause locomotion difficulties in their rear legs. They can be feed pre-frozen pink mice, night crawlers, meal worms, hard-boiled eggs, etc. (Ebenback, 2012, 51). Box turtles are mostly omnivorous and consume night crawlers, earthworms, and meal worms but not wax worms, as they are too fatty. Also, mushrooms, fruits, and small invertebrates such as sow bugs and slugs (Madar, 2005, 274). However, be mindful to purchase store-bought mushrooms and not a backyard variety that could be poisonous.

My box turtles can have protein twice a week. In fact, my adult box turtles would eat any size of night crawlers. I also fed my juvenile red-foot night crawlers once a month. Unfortunately, one of my red-foots developed pyramiding due to the high amount of protein in the night crawlers. "Pyramiding is when carapace scutes grow vertically (like a pyramid) rather horizontally. The appearance of the shell looks bumpy" (Ebenback, 2012, 11). According to Ebenback, it is suggested that giving defrosted turkey burgers to juvenile red-foots avoids pyramiding.

My adult tortoises on the other hand have not been given protein for a while. During my experience, I have never observed locomotion difficulties with my animals. Some people might offer boiled eggs once a month. Unfortunately, my red-foots didn't do well on eggs. It caused diarrhea. I was supposed to take them for a classroom demonstration and had to cancel. It's important to know that too much protein is not good. Balance and variety in their diets is beneficial.

Reptiles with chronic renal disease often have been fed a long-term mismanagement of "high-protein diet rich in purines and in particular the use of canned dog or cat food" (Mader, 2005, 880). Eating these foods can lead to hyperglycemia in herbivores with increased demands for effective urate excretion (Mader, 2005, 880) It is preferred that tortoises and turtles be fed what they find in their natural habitat.

# Plant Protein

"Plant proteins often lack sufficient essential amino acids. Plant species vary in amino acid content. For example, cereals often lack lysine, whereas legumes such as beans and alfalfa often lack methionine. Animal proteins often contain more optimal proportions of amino acids. Protein amounts depend on the right amino acid balance" (Madar, 2005, 257). "Legumes are a good source of calcium and plant protein. Legumes include beans (butter, green, lima, snap, and soy) and peas (snow and sugar). Sprouts may be added to salads. Commercially grown sprouts include legumes (alfalfa, mung bean) and grasses (barley, oats, radish)" (Madar, 2005, 267).

# Plants

"Red-foots will eat vegetation such as clover, violets, irises, and grape leaves." They also "like flowers and will consume rose petals, hyacinths, nasturtiums tiger lilies, coreopsis, marigolds and many other species" (Pingleton, 2009, 87). I also used hibiscus.

# Avoid Certain Plants

Avoid sago palms used as house plants. The nuts (seeds) are extremely toxic and are only produced by female plants. You should bring the chewed plant so the vet may identify the plant species (Madar, 2005, 1077). Also, avoid "oak trees that are found almost worldwide. Acorns, buds, twigs, and leaves have been implicated, but most incidents of intoxication involve either immature leaves in spring or freshly fallen acorns in the spring" (Madar, 2005, 1078). Avoid rhubarb, which is toxic as well (Madar, 2005, 38).

# Mazuri Pellets

I sometimes use the Mazuri pellets soaked and mixed with greens, which lack phytate. Some say you can use the pellets twice a week in various ways. Only research will tell the long-term effects of pelleted diets. However, vets and herpetologists agree that variety for chelonians is the key to healthy pets. Tetra reptile floating food sticks are too hard, even after soaking them. They are a good choice for aquatic species of turtles, such as Red-Eared Sliders. However, they are not for terrestrial turtle species such as box turtles. The box turtle can choke on these as they do not get soft enough even after soaking them. It is preferable to get Mazuri pellets. Soaking pellets and other moist food also add hydration to their diets.

# Calcium

Calcium is an essential nutrient to help the development of healthy bones. Especially for females who are hatching and laying eggs. Deficiency of calcium, vitamin D3, and proper UVB lighting can lead to metabolic bone disease (MBD). Metabolic bone disease can cause malformation of the shell and bones. It will leave the turtles with soft shells and bone deformities such as pyramiding. In the worst cases, their jaws and limbs thicken. If left untreated, this can become life threatening (Ebenhack, 2012, 72).

# Vitamin D3

According to Ebenhack, turtles need vitamin D3 because calcium cannot be absorbed without it. Vitamin D3 "allows the body to maintain a proper balance of calcium and phosphorus." Tortoises do not need vitamin D3 if they are outdoors (2012, 59–61). "UVB light is necessary for metabolism of calcium." When exposed to UVB lighting, vitamin D3 is blended by cells within the skin (Cook, 2008, 50). If turtles don't have the source of ultraviolet light to activate vitamin D3, they can suffer from nutritional deficiencies and die (Flank, 1997, 57).

# Avoid Calcium Injectables for Reptiles

"There is never any justification to treat any pathology with injectable calcium. Aside from being a dangerous practice, calcium injectables have been shown to be painful to the patient and, most importantly, at high doses can cause permanent damage to the patient's kidneys" (Mader, 2005, 846–847). I was ignorant back then about the possible risks involved. I just wanted the shot to help Charley grow. How dangerous that could have been! His weight was carefully measured; he was around 296 grams then.

# Foods That Are Rich in Calcium

Papaya is a wonderful fruit for red-foots and humans alike. "It is rich in calcium, contains enzymes that can aid digestion, and has some antihelminthic (worm-fighting) components as well." The rind, fruit and seed are safe to use for red-foots (Pingleton, 2009, 81). Also, "dandelion leaves are wonderful calcium-rich food for red-foots. Mulberry leaves and ripe fruit are another good food choice" (Pingleton, 2009, 79). Leafy greens are a good source of calcium.

# Phosphorus

When getting bugs, it is preferable not to catch them from the backyard due to possible parasites or pesticides. Catching your own insects is not recommended. I recommend feeding crickets a high-calcium diet for 24 hours or more before feeding them to the herp, which is called *gut-loading*. Gut-loading is the process of feeding crickets or other recommended insects. Gut-loading supplies bought in pet stores are cubes for nutrition and jelly with calcium instead of water, so that crickets will not drown and get contaminated. This feeding is to give good nutrition to the insects. Within 24 hours of feeding the turtles, they ingest the nutrition from the crickets.

Another option would be adding some calcium to the crickets, which is called *dusting*. It is usually done with a Ziploc bag by putting crickets inside the bag with calcium powder. Then, shake it. However, I would instead put a pinch of calcium powder on a moist slice of strawberry to be sure to get the calcium. Because by the time crickets hop all over the tank, much of the powder falls off, especially if they hop out of the turtles' water bowls. Crickets are mostly phosphorus. "Excess phosphorus may cause secondary hyperparathyroidism, bone re-absorption, and calcification of kidney and heart" (Mader, 2005, 287).

A number of years ago, Dr. Jones, told me to watch the number of crickets I fed the turtles. He said that two to six crickets are enough for a small box turtle since crickets contain phosphorus. Strangely enough, Charley got finicky after that conversation, and he turned down crickets for a while.

# Supplementation

A basic diet requires supplementation of vitamins and supplementation and calcium. There are three preferences for calcium. Cutter bones can be used for them to chew on. Also, ultrafine powder calcium with phosphorous-free D3 can be put on food. Please be advised that you should check the directions for the doses. When animals are out in the sun, liquid calcium without D3 can be sprayed on their food. I learned that the old school believed in no calcium or supplementation for adults, but now they believe in supplementing two or three times a week. However, I wouldn't want to do more than once a week for Charley. You should not over-supplement, which can cause dehydration and dangerous bladder stones (Ebenhack, 2012, 73). When doing supplements for hatchling and juvenile turtles, they need to be seen by a vet for the measurements.

As you can see, my book is written to address the problems that can come up with too much of a good thing and at the wrong age group. Giving the right supplements at the right time of the pet's life is essential as a pet grows. You need to know the right amount and how long to give it to your animal. Always ask the vet these things, and be sure you understand what they tell you. If necessary, have them write it down for you. Once the pet is fully grown, you should turn your attention to its diet and ensure it's completely balanced in nutrients, proteins, and vitamins. You need to know how foods interact with each other and with animals. For instance, some vegetables bind minerals; others bind calcium. Attention should be given to what greens and vegetables contain calcium, vitamin A, etc., and what they usually eat in the wild.

# Hypovitaminosis A

Hypovitaminosis A is primarily a disease associated with chelonian box turtles. Indeed, "tortoises are not affected." Vitamin A deficiency caused from being fed an unbalanced diet lacking adequate levels of vitamin A. Certain chelonians, such as box turtles, for example, could "normally have low levels of vitamin A in their livers and retinol in their blood" (Madar, 2005, 831–833). So, they are "more sensitive to dietary deficiencies." However, "most diagnoses of hypovitaminosis A are currently based on dietary history, clinical signs, and their response to treatment. Clinical signs include ocular discharge and blepharedema" (Madar, 2005, 831–833).

# Shut Eye and Cellular Debris

Vitamin A deficiency is called *shut-eye*, as the turtle shuts their eyes. "The eyes are so dry and void of lacrimal secretion that replenishing and maintaining their moisture is crucial." Even though it may appear moist in their eyes (Vosjoli & Kingenberg, 1995, 79), it's recommended that you soak the turtle in a shallow pan with warm water for hours. This will help to get them hydrated.

"Cellular debris under the eyelids" can be removed by using a blunt probe and digital pressure (Madar, 2005, 833). However, this should be administered by a qualified veterinarian.

Turtles with shut-eye will not eat until their eyes are again able to open. If a turtle is "severely underweight, force-feeding or placement of an esophageal-stomach tube is recommended. Box turtles generally respond well to treatment, even in severe cases" (Madar, 2005, 833).

The "long-term treatment and prevention involve expanding the diet to include foods rich in carotenes" (Madar, 2005, 833). "For tortoises and box turtles, these foods include dark leafy greens," spinach in moderation, dandelions that are well rinsed (I buy them from the store), turnip, mustard greens, "yellow or orange colored" fruits or vegetables, steamed winter squash, "carrots, green peppers, sweet potatoes, yams, and cantaloupes" (Madar, 2005, 833). Charley enjoys eating cooked and cooled mashed sweet potatoes. Sweet potatoes are a favorite for all my herps. Yams and sweet potatoes must be baked, mashed, and cooled off. Raw yams are too hard to bite, even for red-foots.

"For aquatic turtles, liver in whole mice or fish is a good source of vitamin A, as are algae and pond weeds. Some commercial treats suitable for herps are Purina Trout Chow and Tetra reptile floating food sticks. These are good sources of vitamin A" (Madar, 2005, 833).

## Vitamin A Injectable

"Vitamin A should not be used unless dietary history and clinical signs suggest a deficiency. Vitamin A deficiency in tortoises remains unlikely given their diet. It should be dosed carefully in other chelonians. Until more is known, avoidance of water-soluble vitamin A and use of fat-soluble vitamin A instead seems prudent" (Madar, 2005, 833).

"Fat-soluble vitamin A is recommended over water-soluble vitamin A….Subcutaneous injections of vitamin A for one or two treatments (every 14 days) is an effective treatment for hypovitaminosis A (shut-eye)….Symptoms usually resolve gradually within two to four weeks, depending on the severity of presenting clinical signs" (Madar, 2005, 833). "Water-soluble vitamin A is much faster acting than fat-soluble vitamin A and perhaps even more toxic. Water-soluble vitamin A produces these lesions within 10 to 15 days, yet fat-soluble vitamin A at identical does not, even though liver levels of vitamin A were similar" (Madar, 2005, 834).

Back in 2008, in Smithtown, New York, the first day I volunteered at the Wildlife Conservation Center, the biologist Eric A. Powers showed me an Eastern box turtle with shut-eye. He told me he was not responding and was no longer eating. I simply said that the turtle had a vitamin A deficiency, so I got him to a vet and gave him a vitamin A injectable. The biologist called me a few days after and said that the box turtle had a rapid recovery after he had the injectable. After that, the biologist and I developed an excellent repose, and he was often glad to take my advice. For injectable or oral vitamin, A, dosages should be administered by a vet.

"Vitamin A is also necessary for normal skin and periocular tissue health, particularly in chelonians with hypovitaminosis A. They typically show ocular discharge, palpebral edema, blindness, hyperkeratosis of skin and mouth parts, and aural abscesses. Patients can improve with vitamin A supplementation: 2000 1U/ kg every 7 days, and better diets" (Madar, 2005, 1069).

"Unfortunately, excessive Iatrogenic administration of vitamin A" can cause in appearance full thickness of skin sloughing, secondary bacterial infections, discoloration of the skin, and extreme lethargy. "These results happen with dosages of 10,000 IU/G or higher IM in a single injection. Treatment involves stopping vitamin A administration, antibiotics, fluid therapy, and nutritional support. The skin lesions may heal slowly but can completely recover. Prognosis varies depending on the severity of the lesions" (Mader 2005, 1069). Avoid too high a dosage of vitamin A or administering it too often, as hypervitaminosis A can occur.

Hypo—too little. Recovery—two to four weeks. Hyper—too much. Recovery—four to six months.

"Treatment of hypervitaminosis A is essentially identical to that of severe burns" (Mader, 2005, 834). Please be advised not to treat hypervitaminosis on your own. A qualified veterinarian can treat this problem.

## Certain Foods to Avoid That Are Toxic

Avoid human food such as grains, bread, pasta, and Bok choy, etc. Raw meat purchased from a store that is not defrosted first can carry bacteria. "Food can serve as a vector for a parasite bacteria, fungus, or virus, so make every effort to avoid wild-caught food, and transfer food from one cage to another" (Madar, 2005, 38).

## Oxalic Acid

Certain veggies that need to be avoided in large amounts are cabbage, broccoli, brussels sprouts, cauliflower, kale, and radishes; these all have been found to contain iodine-binding agents. Also, another group includes spinach, beets, and celery stalks, which contain oxalic acid in sufficient quantities to interfere with normal calcium uptake and metabolism (Mader, 2005, 38) Also avoid potatoes.

# Certain Fruits to Avoid

Some fruits are good for a chelonian's diet in moderation. It will stimulate their appetite and add variety to meals. Some fruits, like melon, can add hydration to their diet and keep them busy. Beware that too many fruits can cause runny stool.

"Reptiles are capable of seeing color. Brightly colored fruits such as strawberries, tomatoes, bananas and melons often attract the attention of many herbivorous lizards and tortoises and invite consumption" (Madar, 2005, 39). Charley is attracted to foods that are orange or red. I also prefer organic strawberries because pesticides stick under strawberry seeds. Try to avoid these seeds. "The seeds of apples, apricots, cherries, peaches, plums, and jetberry bush contain cyanogenic glycosides" (Madar, 2005, 1077). The seeds are dangerous if the seed capsule is broken. "The onset of clinical signs may be very rapid, and death can occur suddenly. Treatment for cyanide toxicosis is often not successful, but does include 100% oxygen administration, supplemented fluids and perhaps sodium nitrate or sodium thiosulfate. Attention must be given to what captive reptiles are eating, being fed or are able to come into contact with" (Madar, 2005, 1077). Also, avoid avocados. "Persin, a compound isolated from the leaves is believed to be toxic and is responsible for the avocado's toxicity" (Madar, 2005, 1077). Avoid eggplants, which are toxic as well (Madar, 2005, 38).

# Diet and Nutritional Needs for Eastern Box Turtles
## (*Terrapene carolina*)

Beetles

Beetle grubs

Blackberries

Blueberries

Cantaloupe

Carrots

Crawfish

Crickets (gut-loaded)

Eggs (whole, uncooked, organic)

Earthworms

Feeder fish (whole, uncooked)

Fish (cooked)

Grapes (seedless)

Grasshoppers

Gutter worms

Hibiscus leaves

Honeydew melons

Lean meat

Mallow leaves

Musk melons

Mice (frozen)

Other berries

Pinkies (newborn, frozen)

Pumpkin

Squash

Shrimp (cooked)

Silkworms

Strawberries

Sweet Potatoes

Turnip greens

Yams

(Madar, 2005, 90)

# Diet and Nutritional Needs for Red-Foot Tortoises
## (*Chelonoids carbonarius*)

Grasses and hays

Green beans

Kale

Kiwi

Leafy greens

Mango

Melon

Mulberry leaves

Mushrooms

Mustard green

Okra

Grasses and hays

Green beans

Kale

Kiwi

Leafy greens

Mango

Melon

Mulberry leaves

Mushrooms

Mustard green

Okra

Papaya

Parsley

Prickle pear cactus pads

Pumpkin

Red-green leaf lettuce

Romaine

Spinach

Sugar snap peas

Tomatoes (no stems or leaves)

Turnip green

Yellow squash

Zucchini squash

(Pingleton, 2009, 79–81)

# Chapter 8

# Correcting Conditions
in Turtle Town

HYLAND

# Humidity

Starting in the summer of 2008, I noticed some problems with Charley's eyes. I kept spraying or misting him, but eventually, he developed dry skin problems around his head, neck, ears, and eyes. I contacted the turtle and tortoise rescuer. She told me that box turtles need more humidity. Soaking alone would not help him, as he returned to his dry enclosure with the heat lamp on over a reptile carpet. This affected his skin and eyes. In the wild, they live in dirt, not in a sterile environment or a tank with a rug in it. They will get too dry; they need to live like wild turtles. They also (especially *Terrapene carolina triunguis*) require a large dish to soak in, where they can wash any substrate out of their eyes. They will be healthier and happier if they do.

I knew I had to take the advice of the turtle rescuer. So after Charley had a good long soak, I let him look at his carpet one last time. I removed the clean green carpet, which had remained clean for over five years. This is because I always soaked Charley each day outside of his tank. I would clean the entire tank with hot water and make the eco-earth from three large blocks. I used hot water to soak eco-earth bricks, then placed Charley in his enclosure. It was nice, warm, and comfortable. Charley was so thrilled with his new environment! He immediately began burrowing like a child rolling in snow for the first time.

He also has a fairly large water dish in the middle of his tank. My orange boy would climb in and out of his reptile pool using the ramp on one end, which I also placed inside his tank. I always kept fresh water in his pool and water dish and misted the eco-earth every morning. I did leave two artificial bushes that hung from the two ends of the tank, one in the front and the other in the back. So, he has a green pool, a green water dish, and two green bushes where he can hide. He can also burrow into the eco-earth by his bush or under the heat lamp. I also placed a heating pad under his tank because, with the humidity high, the temperature drops in the cooler months. I would keep the temperature dial at 85 degrees to low 90s, which can lower or shut it off entirely in the summer.

Sometimes, the humidity would rise too high, even with 50% to 61%. I learned to adjust the humidity to 25% to 40% and kept the temperature on the warm side of a 75-gallon tank. Usually, at 80 to 85 degrees, no more than 90 degrees for winter. Charley would be warm at night, and the same requirements were applied to the red-foot tortoise. However, one exotic vet stated that the humidity should be 50%–60%. I had it too low. After a while, I began to ponder within myself about the logic of humidity.

The substrate, eco-earth, became like dust. What is the point of having high humidity when the eco-earth was drying out and becoming like dust? So, I called Dr. Jones's office, who told me the humidity in my house alone would be around 50% in New York. Currently, I discovered that the humidity was 80% in my house in Florida. So, I went through my textbooks and found a better understanding of humidity.

According to *The Box Turtle Manual*, relative humidity is a critical factor in the successful keeping of box turtles. The relative humidity should be 60% to 80% for *Terrapene carolina* complex. Under low humidity conditions, many box turtles will burrow if given the opportunity to do so. In captivity, box turtles may become stressed if they are unable to burrow when kept at a low relative humidity. They also may develop eye and ear infections under these conditions (Vosjoli & Klingenberg, 1995, 22).

Later, you will learn more about burrowing.

The *Box Turtles* book explains that humidity plays a significant role in benefiting box turtles' growth. "Research has shown that dry conditions contribute to deformities of the shell" (Cook, 2008, 48 and 65). Humidity should be between 70% and 90%. The substrate needs to be sprayed as often as needed. Placing water dishes under lights increases the humidity. Check the temperature and humidity often to ensure they are in the optimum range. If the turtle is not eating, raise the temperature a few degrees, then see if the turtle resumes eating. Misting can also stimulate the appetite, along with new foods and favorite foods (Cook, 2008, 48 and 65).

Another book I read, *Turtles and Tortoises for Dummies*, stated, "The three-toed box turtle enjoys humidity. A cage or enclosure that's too dry can lead to significant health problems, including those of the eyes and the ears" (Palika, 2006, 103). This is why I always make sure Charley's reptile pool is full of fresh, clean water. Palika also tells how the three-toed box turtle enjoys the water. "This turtle is terrestrial; he goes into the water more willingly than the Eastern box turtle does. The three-toed box turtle has even been found foraging in the water. Otherwise, this turtle inhabits a wide variety of terrain, usually staying close to water and food" in their native and natural habitat (2006, 102).

## Enriched Environment

I recently came across more research concerning correcting conditions within inside enclosures. A thesis was written in 2006 by Beth Catherine Case on the "Environmental Enrichment for Captive Eastern Box Turtles" under the direction of Dr. Phil Doerr. Case writes in her abstract:

> Turtles in enclosures had a significantly lower heterophil to lymphocyte ratio (H/L) at the end of the treatment period, indicating they were less stressed than barren housed turtles." Enriched-housed turtles also spent significantly less time in escape behavior, suggesting they were more accepting of their housing conditions....Significant differences in fecal corticosterone or body weight are also different between the two treatments. The test trials proved "turtles showed a distinct preference for the enriched environment."...This study shows that the captive housing environment can negatively or positively influence the physiology and behavior of box turtles. Housing modifications that encourage typical species behavior should be provided for the box turtle. These would include substrate in which to dig and items that permit hiding. (Abstract)

> "The enriched environment, comprising the other half (39 x 35 cm) of the enclosure, was covered with 2–4 cm of cypress mulch" (Case, 2003, 25).

"Another logical starting point for an enrichment program is the physical space of the enclosure... cage design, and furnishings all must facilitate an array of species-specific behaviors such as locomotion, physical postures, play, and exploration" (Case, 2003, 4).

"There are numerous physiological measures of stress and poor welfare. These include anorexia, weight loss, heart rate changes, immunosuppression, changes in hormone production and reproduction impairment" (Case, 2003, 5).

"Chronic stress may be more accurately monitored by assessing immune function, as measured by changes in leukocyte count (Gross and Siegel 1983). Cellular changes associated with stress-leukograms are an increase in heterophils (equivalent to mammalian heterophils), and a decrease in lymphocytes" (Case, 2003, 6).

"The impact of housing on immune function, which is reflected by the difference in H/L ratios, has important implications for turtles in long-term captivity. For turtles kept as pets or in educational exhibits, housing conditions may ultimately influence health and longevity" (Case, 2003, 46).

"There is both physiological and behavioral evidence that turtles benefit from an enriched environment. The substantially decreased H/L ratio observed in the enriched-housed turtles indicates they experienced less stress than barren-housed turtles. Similarly, the marked difference in activity budgets, especially escape attempts, suggests the enriched-housed turtles were more accepting of their housing conditions" (Case, 2003, 45).

Also, "turtles living in enriched enclosures spent greater amounts of time resting and performing no escape movement relative to barren-housed turtles. Barren-housed turtles spent more time engaged in escape behavior than enriched-housed turtles" (Case, 2003, 44–45). Enclosure enrichment for captive box turtles "enables the turtles to perform natural behaviors, such as digging and hiding" (Case, 2003, 47).

"Hiding materials are an important aspect of the box turtle's environment. When available, these hiding places were utilized 91% of the time, and even when not made available, turtles often create their own hiding places" (Case, 2003, 47).

Before the eco-earth, I would have a hide box with artificial leaves and bushes in Charley's enclosure. I kept what was good in his enclosure and added what was better: the natural eco-earth and bushes. That way, he could hide and look around while he dug through his leaves from the hanging bushes or hide in a washcloth in his enclosure. (See chapter 5).

# Burrow

As explained in "Environmental Enrichment for Captive Eastern Box Turtles," box turtles in the wild "often burrow into leafy debris, rotten logs, brush, and soil.…These shallow burrows are called resting forms and may function to regulate temperature and hydration or to avoid detection" (Case, 38–39). Turtles can hide in these forms for days or weeks, even if conditions are unfavorable for travel (Case, 38–39). These results from various studies show surprisingly that the *Terrapene carolina carolina* (Eastern box turtle) or *Terrapene carolina triunguis* (three-toed turtle) naturally spend most of their time burrowing. This is its normal behavior in the wild for enriched housing (Case, 38–39). In the wild, the Eastern "box turtle is terrestrial and typically inhabits open wetlands but may also be found in marshy meadows or grasslands" (Case, 8).

Charley would burrow in the eco-earth about five inches deep and could stay there a long time. While burrowed, he would stick his head out and stretch his long neck. Dr. Jones said some research showed if turtles stick out their heads while burrowing, they still get the benefits of UVB lighting.

Peekaboo!!!

## Temperature Gradient or Thermal Gradient

Different ranges of temperature may vary with different species such as box turtles and tortoises. In research, according to the authors, "All animals have a set point temperature or a set temperature range regulated by the hypothalamus, a region of the brain that controls temperature regulation" (Zug et al., 2001, 179). "The set point temperature is essentially the thermostat setting that signals when an animal should initiate body temperature regulation" (Zug et al., 2001, 179). In ectotherms, this "response is usually behavioral and, to a lesser degree, physiological....As the ectotherms body temperature shifts away from the set point temperature, the animal will move, change its position, or change its orientation" (Zug et al., 2001, 179). In this case, getting a thermal gradient and enough room is essential. It will allow them to have space by moving from warmer or cooler areas in the enclosures. They can move around and regulate their body temperature (Flank, 1997, 54–55).

So, I put dials around the inside of the tank for both temperature and humidity for the box turtles and tortoises. The temperature control dial is under Charley's tank in cooler months, such as fall and spring.

This device is great for shutting off heat lamps when I am not at home. Unfortunately, watts over 100, especially the ceramic heat emitters, blow out. I realized that the wattage was too high for the temperature control.

Even though my book is mainly about box turtles, I will add some information about this other charming chelonian, the red-footed tortoise. They are some of the most beautiful tortoises in their coloring and patterns—and one of the most personable and friendly!

I consider all my chelonians sweet and docile. The red-footed tortoise is similar to the yellow-footed tortoise but is a bit larger species than the yellow-foots. Both originated from South America, making them highly tropical. They cannot tolerate cool temperatures of 60 degrees or less for two days but like to be in temperatures ranging from 80 to 85 degrees, with a short cool down at 65 degrees.

Rosie: below

Edwina aka Eddy: below

Jeanne Hyland

# UVB Lighting (Ultraviolet B)

UVB 5.0 tubes are recommended for tropical species. Never use a 7.0 UVB because it can cause renal damage in tropical species. The 7.0 should only be used for desert species. UVB 5.0 radiations are used for tropical species, such as Eastern box turtles, which burrow in leaves and do not receive direct sunlight all day, only sunlight that is filtered through the trees appears in their natural environment. Red-foot tortoises live under the canopy in a tropical rainforest. Their sunlight is also filtered through the thick forest canopy.

Compact screw-on UVB lighting can cause eye damage to chelonians. I was informed by Dr. Jones and a turtle rescuer never to use the compact UVB twist-on bulb for turtles. They recommended using the long fluorescent tube bulbs with only a 5.0 tropical bulb. I took the compact bulbs off immediately and purchased the tube lighting. Both turtles had eye issues at that time in the two tanks that had compact screw-on bulbs. Thanks to the advice of Dr. Jones and a turtle rescuer, both turtles stopped having eye problems when I began to use the correct lights.

# Hibernation/Brumation Tropical Box Turtles

Tropical box turtles or those from the southeastern U.S., do not require as intense a hibernation period. Those from the Southeast thrive after being hibernated for only four to six weeks. Tropical turtles usually undergo a period of inactivity that coincides with the dry season in the areas from which they originate. This period of rest likely has an important influence on their hormonal metabolism, so it is advisable to try to duplicate this situation. To simulate this period simply reduce light intensity and diminish feedings for about six weeks. (Vosjoli & Klingenberg, 1995, 99)

You do not need to reduce temperatures except through shorter exposure to a basking light (Vosjoli & Klingenberg, 1997, 99). However, this is not recommended for:

1. Sick turtles.

2. Dehydrated and malnourished turtles.

3. Turtles who are underweight (because they must live off their own body resources during hibernation).

4. They must stop feeding at least ten days before hibernating; remaining food left in the upper intestinal tract can lead to infection and toxicity. Once the infection or other health concerns have been addressed, allow the turtle to hibernate for a short period, four to six weeks (Vosjoli & Klingenberg, 1995, 98).

If box turtles are dehydrated during hibernation, one good idea is to place hibernating box turtles in a plastic sweater storage box. That would have four to five inches of lightly moistened, damp but not wet substrate, using newspaper strips and peat moss. Ensure the sweater box has just enough holes for air passage but not enough to allow the contents to dry (Vosjoli & Klingenberg, 1995, 99). Check to ensure the substrate is moist enough every two to three weeks.

In the meantime, "allow the turtles to soak in a shallow container for one to two hours," then place them back into the container (Vosjoli & Klingenberg, 1995, 99). Examine the turtle for illness, shut eyes, difficulty breathing, and any bubbling from the nose. If you find any ailments, remove the turtle from the container and place it in its normal husbandry conditions, taking steps to remedy all ailments. Healthy turtles can easily hibernate for three to four months (Vosjoli & Klingenberg, 1995, 99).

The difference between hibernation and brumation is that when a bear hibernates in a cave, their body temperature does not drop. Only the environment around them drops, unlike reptiles whose body temperature drops, like the environment around them. This is called brumation, not hibernation. Turtles should be under the frost line, insulated by leaves.

# Results of Raising Humidity

After corresponding with a turtle rescuer to learn more about raising box turtles and red-foots, I was able to enrich Charley's enclosure and keep up with the humidity. In the summer months, I increase the humidity. In the winter, I regulate enough heat by using overhead heating (ceramic heat emitters) to balance out the normal humidity and not let it get too cold.

I try to keep in the preferred optimum temperature range (POTR), which is between 85 and 88 degrees on the warm side of the tank or enclosure and 70 and 75 degrees on the cool side of the enclosure.

I discovered that some research shows that three-toed box turtles slowly lose their juvenile marking as they reach maturity. However, raising the humidity helps turtles to keep their bright colors. I notice Charley's brighter colors on his skin instead of a dull grey-orange color. My housemate Heidi has known Charley for over two years. She mentioned to me that she had also noticed his color. Charley would be soaked every morning and eat twice a week. I make sure I mist his tank daily with lukewarm water since I have made changes in the humidity. I see a big difference in Charley, my handsome herp!

I also changed and corrected the conditions by adding cypress mulch. Cypress mulch is much better as a substrate as it holds humidity and is mold resistant. Compared to eco-earth's coconut fiber, which becomes dusty, dries out, and gets into chelonians' eyes. According to a volunteer from the Chelonian Research Institute, box turtles need semi-swampy humidity and red-foots need high humidity but not swampy conditions. Unfortunately, after trying semi-swampy conditions, it has led to mildew that causes shell rot, which has taken me months to undo using topical treatment. Drying it out overnight is the only way to raise the humidity inside. You can use your judgment on what is preferred for your turtles. I have found that my house's humidity varies from 75% to 80%. Box turtles that live in the wild inhabit "moist forested areas but also wet meadows, pastures, and floodplains" (Behler & King, 1979, 468). So, it's important to make them feel at home and comfortable.

## Husbandry and Housing Hints

Box turtles avoid stress by:

1. Having hiding places such as hide boxes, logs, brushes, and burrows.
   Avoiding toxic plants. I recommend artificial plants made just for reptiles.

2. Having the right substrate. Anything you put on the bottom of a tank or enclosure for a deep, moist burrow, such as eco-earth, also called coconut fiber. I changed to cypress mulch, which works well for misting daily with water spray bottles. Keep in mind that the humidity hydration helps skin and eyes as well. Stick temperature and humidity dials, two on each end, inside the tank.

3. Soaking. They have to hydrate in soaks. Use shallow water like plastic trays. Shallow water is essential so the turtles don't drown. Make sure to watch them so that they don't flip upside down in their water.

4. In a 75 to 100-gallon tank, you need two places to burrow: one on the warm side by the heat lamp and the other on the cool side. Note: Don't put the heat on in the warm months. Use a temperature control, which shuts off heat lamps if you're away from home.

5. Use long-lasting ceramic infrared heat emitters. All UVB lighting tubes should be replaced every six months for young adults. However, I usually replaced them once a year.

6. Large enclosures containing several turtles should be cleaned every two to three months, and smaller tanks should also be cleaned every two to three months. You can also do spot cleaning between changes.

# Chapter 9
# Celie's Homecoming

HYLAND

On September 9, 2015, I received a message on my phone from a vet tech, who said, "I have found a girl for Charlie." I returned his call and told him, "I will be coming over to see her." In the office lobby of the animal clinic, I saw her. The yellow spots on her brown face and her sweet brown eyes drew my attention, which matched the yellow and tan scales on her legs.

Celie's oval-shaped shell was so smooth that it felt like a polished wood surface. After thanking the vet tech "from the bottom of my heart," I took her home. A few years ago, I found out she is a very, very old turtle. Celie is sweet and shy. She would either hide in her shell or walk away; she doesn't bite as the Florida box turtle does. I have decided to clean out Edwina's (the red-foot tortoise) old 75-gallon tank and sanitize it with Nolvasan, then fill the tank with eco-earth. Soon, she will meet Charley.

Below is Celie.

# Chapter 10
# Charley's 35th Herp Birthday
## October 18, 2015

HYLAND

I celebrated my chelonian Charley's 35th birthday. He has been with me for 30 years. I took pictures all day of him cherishing the years with me. Charley has given me nothing but joy and pleasure. He got so much attention on that day. I put a huge light orange bow on the outside wall of his tank. Then I fed him freshly cooked, mashed orange yams, the same color as he is. Also, Eddie and Rosie were fed fresh yams while they were joining the chelonian celebration.

Charley loves attention. I even sang "Happy Birthday" to him as he ate his sweet potatoes. "Boy, he sure loves those sweet potatoes!" Then, I held him a few times. Even at 35 years old, Charley is quite active and healthy and can be a little spoiled. He begs by standing on his hind legs, which he has done five times a week. Sometimes, I give him more crickets when Charley begs. Why, he ate four lively crickets after begging the other day! Other times, I will provide him with strawberries for begging. Charley is still small, and he has a hard time holding the crickets. So, I still hand-feed him. In the following chapter, Charley finds his special gift, which is Celie.

**Florida**

# Chapter 11
# Reptile Romance

HYLAND

Charley is so excited to see his girlfriend, Celie. He gets busy with her for over three hours until he falls off. After Celie is returned to her old tank, Charley goes looking for her, pecking on the glass with his little orange nose. "Aren't they a good pair?"

# Chapter 12

# Charley's 42nd Herp Birthday
## October 18, 2022

HYLAND

On this special day, I took out my camera and started taking pictures of my little guy, Charley. While I was clicking away, Charley sat at the corner of the tank and watched me. It's almost like he was posing for the camera! So I took him out for a soak alongside Celie in separate pans. After the soak, I took more pictures of him eating and enjoying himself.

I also photographed him sitting under his big orange bow with the "Happy Birthday" letters in turquoise. That evening, my little boy Charley looked so proud, his head held up high as a happy herp. He has been a close pet and a lifelong companion, with me for most of my adult life. Sometimes, the best, most valuable gifts come in small packages.

# Chapter 13
# Turtle Intelligence

HYLAND

For over 40 years I have observed Charley, my box turtle. Within this time I have noticed more intelligence than most people realize turtles have. They can be even trained to learn their names. Charley was on the glass in his tank several times, facing the opposite end. When I came in and started to call "Charley, Charley. Hello, Charley," etc., he practically ran over to me to look out of the tank with his head held up high so he could see me.

Charley always knows just how to communicate with me when he wants something. When I would keep mulch in the corner of his tank, he would push the mulch away from the corner of his tank and hit his bottom shell on the glass to make a clunking noise. By doing this, Charley got my attention and alerted me that he must go to the bathroom. Then, he'd stop clunking on the glass as soon as I came to him. He'd look right up at me, knowing I have come to help him. Sometimes Charley scratches the glass, standing on his hind legs, stops, and looks at me when I approach him.

Another time is when I hold Charley and pet his paws. He wraps his paws around my fingers, and I tell him, "Charley, you're a good boy!" and talk to him. He stretches his head all the way out to my face and puts his nose on my nose—almost like a kiss!

"This is a cute moment that I will never forget."

Charley is very inquisitive about his surroundings. Even from the first, he walked over to my mother in the pet store tank. Last year, when I brought him home from boarding at the vet, I placed him in his tank. I noticed Charley's head stretched out, leaning against the glass, looking for his friend Edwina, the red-foot tortoise in the adjacent tank. Then he looked on the other side of the tank, toward Rosie's tank. A little while later, he was so motionless with a scared look in his eyes, the same look he gets when I take him to the vet's office. He was pretty upset when he discovered that his friends were missing. After the red-foots came home, Charley returned to his usual happy self. He would watch them a few times by scratching the glass again and rocking his head side to side, excited to see them again. Charley would even watch them while he was burrowing, going so far as cleaning off the eco-earth from the tank's wall. Another time was when I took Celie and Charleen out of their tank to clean it. For the next several hours, Charley was pacing in his tank and scratching the glass. After placing the girls back in their tank, Charley quietly went over to his corner in the tank and relaxed for the rest of the day.

I realize that turtles have emotions and feelings on their own level. Someone once said to me in Long Island, New York that, "Charley recognizes your voice from down the hallway." He would move his head whenever his name was mentioned. On another occasion, I had met a friend named Kaycee Clancy at my congregation. We would get into conversations about my turtles, especially Charley. Kaycee grew more curious to see this turtle, and one day I asked her to visit. In Kaycee's words, "she introduced me to all of her turtles, and they just did what turtles do. Then she showed me Charley by calling his name. Charley immediately lifted up his head and looked in our direction. I started talking to him and calling his name, and the more I spoke, the more he stretched his neck and showed his beautiful colors to me. I was very delighted that a turtle would respond like that."

There was also an incident when I was vacuuming the carpet by the tanks in the pet room. Suddenly, the canister opened up, and the dirt went all over the carpet. I yelled out and had a hissy fit. I noticed Charley sticking out his head from his burrow to see what was wrong. Other tortoises were pacing up and down in their tanks. They all respond to the tone of my voice, whether I am calm or upset.

Chelonians can be responsive, friendly, and sweet. A while ago, Charley walked over in his tank to a guest who came to my house. The man was amazed that the turtle responded to him. I told him that Charley was coming over to see him and greet the guest. Charley recognizes when there is somebody new in his room. When strangers enter the tortoises' room, sometimes they sniff up heavily, with their throats going in and out more noticeably.

Charley can be very smart and clever. I emptied his green ramp pool and filled it with paper towels. To make him a little bed to burrow in. Charley immediately walked right over to the pool and into his bed. The next day, Charley was restless, so I emptied his green reptile ramp pool and again filled it with paper towels. He came right over to it and got into the bed. You see, he remembered the days when I kept a bed for him in his tank. I also observed Charley scratching the temperature dial, even trying to bite it so it would move out of his view. I solved that problem by moving the dial up higher.

I often sense that people tend to put a greater value on larger domestic animals. To them, a turtle or tortoise seems like a small, insufficient animal! However, in the ecosystem, each living species has value and worth. The extinction of any species creates an imbalance in the delicate ecosystem, given the community within the ecosystem. (Note: A species is a population within a community or ecosystem.)

Photo by Melissa Podlesney Maravell

# Herp Humor

How do you take a turtle's temperature? You don't…
Turtles are ectothermic and need an outside heat source.

Do turtles get dental care? Yes. By only rubbing a hard, bony beak.

**Laughter is the best medicine!**

# Chapter 14

## Courage to the Challenge

HYLAND

This book contains the life story of Charley the chelonian. Events are highlighted, such as his rehabilitation and progress, including changing the conditions that caused some of the problems in the first place; it has been 19 years since he was rehabilitated. Charley is now thriving at a good old age of 44. My hope is that those who want to learn more about turtle care will be inspired and encouraged not to give up on any challenge.

**A reminder: Have an exotic vet who specializes in reptiles, and who has a good reference from a turtle rescuer and herpetology society.**

# Chapter 15

# Dreams Determine Destination

HYLAND

Every creature has value and worth, no matter its size. Charley has survived so many would-be accidents and has recovered from four life-threatening illnesses in one year. These stories have a purpose in his turtle testimony. Charley has been narrowly spared over and over again, and his recovery has inspired me to study herpetology. As in adversity and trials when my animals were sick, I had the chance to learn herpetology and had the opportunity to volunteer in a wildlife facility at Caleb Smith Park in Smithtown, New York. I cared for the box turtles and red-foots in the conservation program.

I know each creature is unique in this world, and like fingerprints, they are not exactly alike. Charley is a special turtle, one of a kind! During the eleven months I was taking him to the vet till his recovery, I remember one particular office visit. A woman who came frequently commented to me, "Oh, you live here too." My mom and I exhausted the care credit card and put more on the Visa.

I even took him to five vets that year, in 2003. The last visit to the 24-hour hospital, where I dropped him off for four weeks, cost me $1,300. I never regretted helping my boy, Charley. It's incredible how one little turtle changed my life's direction.

Charley the chelonian's turtle testimony is that you have real value and worth no matter how small or insignificant you may feel. Charley is an example to us, showing us that, however small we may be or however small we may feel, we each have a plan and a purpose for our life here on earth.

Perhaps you have been told that you are nothing and will never accomplish anything of worth, but that is not true. Once you realize that you have a purpose, reason, and plan for your life, the possibilities become endless, and the sky is the limit when your dreams become your destination. Through the hard times, through danger and trials, we hide in our "shell," protecting our soft and tender parts within our armor. We protect our dreams from discouraging words. If we can make our dreams determine our destination, then each of us can be a champion like Charley! We take on the challenge given to us in Aesop's fable, "The Tortoise and The Hare." The tortoise was slow, but he won the race. So, take a tip from Charley the chelonian and let your dreams determine your destination! **Hold your head up high!**

**Thank you very much for reading my book.
I hope that you have enjoyed it and that it helps and inspires you.**

# About the Author

Jeanne Hyland enjoys understanding and caring for turtles. She has been taking care of these animals for over 50 years. When she first laid her eyes on these beautiful creatures, she just fell in love. From 1995 to 1997, Jeanne volunteered for the Suffolk BOCES Outdoor Environmental Education Program in Herpetology at Caleb Smith State Park on Long Island under biologist Eric A. Powers. She also got the opportunity to volunteer at the Chelonian Research Institute in Oviedo, Florida, from 2015 to 2018. The Chelonian Research Institute was a nonprofit research and conservation of turtles and tortoises worldwide. It was founded by Dr. Peter Pritchard in 1997. Unfortunately, due to his passing, the Chelonian Research Institute has closed down. Working with the Pritchards, was an honor. They were so generous that they invited her to dinners with them.

J. Hyland completed a course in animal exotics, herpetology, and veterinary assistant. She earned her master's in art education, MS, at C. W. Post, Long Island University in Brookville/Greenvale, New York, along with art certification for nursery and K–12. From 1982 to 1984, Ms. Hyland was a substitute teacher for all subjects in K–12 in Suffolk County, New York.

Jeanne was an art teacher's aide for Unity Drive Elementary School, Centereach, Long Island, New York, in 1980. She was also an art therapist from 1978 to 1979 at Central Islip Psychiatric Center in Central Islip, Long Island, New York.

In 1992, Jeanne completed two years of College Hebrew at Stony Brook University in Long Island, New York. She was a teacher's assistant for the Hebrew Class at Beth Emanuel fellowship. Currently, she attends Beth Israel Synagogue and reads in Hebrew during services. J. Hyland had the opportunity to travel overseas to Europe twice (France in 1971 and Germany, Austria, and Czechoslovakia in 1990) and Israel twice, in 1984 and 1988.

In addition to taking care of her babies, Jeanne enjoys drawing, painting, and reading Hebrew in her spare time. She found faith in the good shepherd who has guided her all her life according to the scripture of the Bible, especially the Gospel of John, chapter 10, and Isaiah, chapter 53, which referred to the Messiah

Yeshua, a.k.a. Jesus. Please read these two chapters.

Her hopes and dreams are to save and protect the lives of these animals when they are kept as pets. She believes it's essential to educate yourself before purchasing a turtle and to always remember: "A turtle is not a toy."

**Science Center (Caleb Smith Park, Smithtown)**

**Long Island New York**

**Mitzi**

**She was a friend of Peggy and Becky that played together.**

**Author Jeanne Hyland**

# References

Bartlett, R. D., & Bartlett, P. (2001). *Box turtles.* Barron's Educational Series.

Behler, J. L., & King, F. W. (1979). *Field guide to North American reptiles and amphibians.* Knopf.

Case, B. C. (2003, July 9). *Environmental enrichment for captive eastern box turtles (Terrapene carolina carolina).* North Carolina State University Libraries. https://repository.lib.ncsu.edu/items/bbb217d7-027e-4392-b789-8af339a42612

Cook, T. (2008). *Box turtles: A complete guide to terrapene and cuora,* T. F. H. Publications.

Dodd, C. K. (2001). *North American box turtles: A natural history.* University of Oklahoma Press.

Ebenhack, A. *Red-footed tortoises in captivity.* ECO Herpetological Publishing & Distribution, 2012.

Flank, L. (1997). *The turtle.* Howell Book House.

Mader, D. R. (2005) *Reptile medicine and surgery* (second ed.). Saunders Elsevier.

Palika, L. (2001). *Turtles and tortoises for dummies.* For Dummies.

Patterson, J. (1994). *The guide to owning a red-eared slider.* T. F. H. Publications.

Pingleton, M. (2009). *The redfoot manual: A beginner's guide to the redfoot tortoise.* Art Gecko.

Vosjoli, P. D. (1997). *General care and maintenance of popular tortoises.* Advanced Vivarium Systems.

Vosjoli, P. D., & Klingenberg, R. J. (1995). *The box turtle manual.* Advanced Vivarium Systems.

Walls, J. G. (1996). *Tortoises: Natural history, care and breeding in captivity.* T. F. H. Publications.

Zug, G. R., Vitt, L. J., & Caldwell, J. P. (2001). *Herpetology: An introductory biology of amphibians and reptiles* (second ed.). Academic Press.

## Web Resources

Box Turtle Site. (n.d.). https://boxturtlesite.info/box-turtle-anatom

Chelonian Research Foundation. (2007). https://chelonian.org/

The Redfoot Tortoise (Chelonoidis carbonaria). (2014). http://redfoottortoise.com/redfoot_tortoise_diet.htm

Turtle Rescue of Long Island. (2018). http://www.turtlerescues.org/

# Further Reading and Nuggets of Knowledge

1.  *The Complete North American Box Turtle* by Carl J. Franklin and David C. Killpack. 2009. ECO/Serpents Tale NHBD.

2.  *Medicine and Surgery of Tortoises and Turtles*, 1st ed., by Stuart McArthur, Roger Wilkinson, and Jean Meyer. 2004. Wiley-Blackwell.

3.  *Encyclopedia of Turtles,* 1st ed., by Dr. Peter C. H. Pritchard. 1979. T. F. H. Publications.

4.  *Living Turtles of The World,* 1st ed., by Dr. Peter C. H. Pritchard. 1967. T. F. H. Publications.

5.  *The Galapagos Tortoises: Nomenclatural and Survival Status (Chelonian Research Monographs)* by Dr. Peter C. H. Pritchard. 1996. Chelonian Research Foundation.

6.  *Sea Turtles of the Eastern Pacific: Advances in Research and Conservation (Arizona- Sonora Desert Museum Studies in Natural History)*, 1st ed., by Dr. Peter C. H. Pritchard (foreword), Jeffrey A. Semioff (editor), and Bryan P. Wallace (editor). 2012. University of Arizona Press.

7.  *Evolutionary Significant Units Versus Geopolitical Taxonomy: Molecular Systematics of an Endangered Sea Turtle (Genus Chelonia) and Status of the Black Turtle* by Peter C. H. Karl, Stephen A. Bowen, Brian W. Bowen. 1999. http://seaturtle.org/library/KarlSA_1999_ConservBiol.pdf

8.  *Rafetus, The Curve of Extinction: The Story of the Giant Softshell Turtle of the Yangtze and Red Rivers* by Dr. Peter C. H. Pritchard. 2012. Living Art Publishing.

North American box turtles are protected because they are an endangered species. They are no longer sold in stores. However, a person can adopt a turtle from a rescuer. Turtles that are injured or sick and those that have been kept in captivity cannot be released in the wild. It is cruel to release a turtle that is kept in captivity for a while since they will not survive. Sick and injured turtles have to be treated by a rescuer or a veterinarian.

I support TSA, Turtle Survival Alliance, which is a well-known nonprofit conservation organization. They treat sick and injured animals and breed endangered species. When breeding endangered species, Turtle Survival Alliance releases the turtles into the wild where they can find a healthy ecosystem that has a large growing population of other turtles of the same species. TSA had many successes and was supported by Dr. Peter Pritchard.

Much prayer and faith have been the glue in putting this book together. I had to face many obstacles, but the Lord brought just the right people into my path in organizing the book to bring it to fruition. When Charley was sick in 2003, it took much prayer over those long months. I thank the Lord for his recovery. One of my favorite verses in the scripture is from Psalms 36:5–6: "Your love Lord reaches to the heavens, your faithfulness to the skies. You preserve people and animals." Another scripture that I like to read is "The righteous person regards the life of his animals." Proverbs 12:10.

www.ingramcontent.com/pod-product-compliance
Lightning Source LLC
Chambersburg PA
CBHW041950220326
41599CB00004BA/145